前沿科学的交叉与融合

——留法学人跨学科研究论文集（二）

Frontier Science: Interdisciplinarity and Integration

ASICEF Workshop on Interdisciplinary Research Topics Ⅱ

主　编　张勇民

副主编　贾子先　陈　勇　谢小燕

中国海洋大学出版社

·青岛·

图书在版编目（CIP）数据

前沿科学的交叉与融合：留法学人跨学科研究论文集. 二：汉、法／张勇民主编. —青岛：中国海洋大学出版社，2019.1

ISBN 978-7-5670-2089-4

Ⅰ.①前… Ⅱ.①张… Ⅲ.①跨学科学—研究—汉、法 Ⅳ.①G301

中国版本图书馆CIP数据核字（2019）第020142号

出版发行	中国海洋大学出版社
社　　址	青岛市香港东路23号　　**邮政编码**　266071
网　　址	http://pub.ouc.edu.cn
出 版 人	杨立敏
责任编辑	矫恒鹏
电　　话	0532-85902349
电子信箱	2586345806@qq.com
印　　制	日照报业印刷有限公司
版　　次	2019年5月第1版
印　　次	2019年5月第1次印刷
成品尺寸	170 mm × 230 mm
印　　张	8.5
字　　数	110千
印　　数	1～1000
定　　价	46.00元
订购电话	0532-82032573（传真）

发现印装质量问题，请致电0633-8221365，由印刷厂负责调换。

前言

在人类发展的历史长河中，科学技术给人们的生活带来了巨大的改变，极大地提高了生活质量，丰富了生活内容。古代科技的发明，将人类由野蛮带入文明，近代自然科学的诞生和产业技术革命的兴起，使人类从农业文明社会迈入工业文明社会。随着现代科技的迅猛发展，科学技术在世界经济社会发展中所起的主导作用也越来越显著。然而，当代科学的发展和重大科学技术成就的取得已经越来越依赖于不同学科间的交叉与融合，许多有影响的科技成果都是在学科的交叉点上取得的。

为了促进不同学科领域学者之间的交流与合作，以便发现新的学科研究交叉点，全法中国科技工作者协会自2006年起，在法国举办了多场跨学科研讨会，产生了很好的效果。2012年，以全法中国科技工作者协会的六位资深学者为主体，在江汉大学成立了交叉学科研究院，旨在追踪学科前沿，促进多学科交叉创新，凝练学科特色，建立高水平的跨学科研究平台。为广泛宣传、全面介绍法国华人学者在交叉学科领域的学术交流和科研探索，全法中国科技工作者协会曾于2008年整理出版了《前沿科学的交叉与融合——留法学人跨学科研究论文集》（北京大学出版社），受到读者的欢迎。现在，我们编纂的第二本论文集即将出版，我代表编委会对中国海洋大学出版社在本书出版过程中给予的支持和帮助表示衷心的感谢，同时感谢论文集的作者，谢谢他们辛勤的撰稿工作。

本书共收集了14篇来自不同研究领域的旅法中国学者撰写的文章，内容涉及数学、物理学、化学、生物学、医学、表观遗传学、纳米科技、汽车发动机、机器人、环境保护等科学技术领域。希望读者能从中了解和学习相关学科的研究前沿和最新进展。

张勇民

2018年11月21日于巴黎

附：

全法中国科技工作者协会（简称全法中国科协）成立于1992年，由来自中国在法国工作的科技工作者组成。全法中国科协的主要任务是团结中国在法科技工作者，组织科技、文娱、生活等领域的交流活动，增进友谊；在促进中法两国的教育与科技合作的活动中，发挥纽带与桥梁的作用。全法科协现有会员二百余人，汇聚了留法的科技精英。许多会员是法国知名大学的教授、国家科研中心主任研究员、企业家或高级工程师。他们活跃在各自的科研和工程技术领域，有些已成为该领域的领军人物，得到中法两国科技界的普遍认可。协会会员与国内同行广泛合作，许多会员受聘为长江学者讲座教授、中国科学院海外评审专家，长江学者海外评审专家以及国家自然科学基金委海外评审专家等。协会会员积极参与中法联合研究生的培养、中法联合实验室的建设及各项中法合作研究计划的工作，得到两国政府的大力支持和资助。协会在中国驻法使馆教育处和教育部“春晖计划”的支持下，积极开展回国服务特别是支援西部的工作，多次组团到甘肃、贵州、陕西、四川、云南、宁夏等地，为那里的科技教育事业添砖加瓦。此外，协会还与国内外有关团社建立了广泛的联系，为发挥协会会员的积极作用提供更广阔的空间。

协会网页：https://asicef.fr/

目录

微纳制造、微流控及生物介观仿真

陈 勇

细胞生物学通常需要对“取出动物或人类细胞在体外进行扩增及各种操作，而这些扩增及操作一般会在与体内完全不同的条件下进行”。简单地说，细胞在体内所处的环境是三维和动态的，而培养皿中所处的环境是二维和静态的。因此有必要对体内的细胞微环境进行仿真，包括细胞外基质的重构、细胞周围可溶性因子的调控、细胞与细胞相互作用和其它物理量的模拟等，以利更好地研究和调控细胞与细胞微环境的相互作用。这可以说是生物介观仿真的主要内容和目的。

现代生命科学使我们对生物分子、基因、细胞及组织器官的结构和功能有了深入的了解，相对而言，我们对细胞微环境重要性的认识还十分有限。近年来，人们对诱导多能干细胞（induced pluripotent stem cell）在再生医学方面的应用寄予了很大的希望，但由于这类细胞对环境的特殊依赖性，其扩增和分化的调控很难在现有的细胞培养体系中完成，从而在一定程度上限制了再生医学的发展。所以说有必要尽快建立以仿真为基础的细胞微环境模型和生产方法。

生物介观仿真是一个前沿交叉课题，它的基础是细胞生物学、生物材料学及微纳制造和微流控技术。就微纳制造技术而言，此前还很少从生物应用这一角度来考虑，现在还没有一种好的方法制造生物材料的图案化微纳结构。就微流控技术而言，由于多样性和工艺的非标准化，其应用还十分有限。因此，推广微纳制造及微流控技术在生命科学中的应用不仅意义重大，而且具有很强的挑战性。本文旨在简要介绍微纳制造和微流控技术及目前作者实验室进行的几个研究课题。

一、微纳制造技术

微纳制造的发展可以追溯到20世纪50年代。当时为制造半导体集成电路开发了一整套芯片图形复制和转移的技术。随着社会经济的发展和社会需求的增加，半导体芯片技术特别是其核心工艺，即光学曝光和刻蚀技术也得到了不断地发展。目前的半导体集成电路加工工艺以深紫外光光刻（193纳米光源）技术为基础，下一代芯片制造将以极限紫外光刻技术（13纳米光源）为基础。除了极限紫外光刻技术以外，还曾发展了诸如X光光刻、电子或离子束投影光刻、微型电子束阵列光刻等高分辨率刻蚀技术。一般来说，这些技术由于设备及应用成本过高且工艺过于复杂，均不适合用于非半导体器件的生产。

相比之下，非传统的微纳制造技术在近些年得到了广泛的重视和发展。这些技术的特点是方法简单、制作成本低，应用面广。例如基于模压技术的纳米压印技术（nanoimprint lithography）在非半导体领域的应用已得到广泛的认可。 纳米压印是通过加温和加压将模板上的纳米图案复制到涂在基片上的聚合物薄膜层上，然后再对其进行刻蚀或剥离处理以得到所需的纳米结构。这一技术不仅复制分辨率和产额高、工艺简单，而且成本低，因此具有很强的竞争力和广阔的应用前景。这一技术的不足之处是需要高温和高压，不利于图形的精确定位与套刻。早些年，我们还提出了紫外光纳米压印技术：先在基片上旋涂一层流动性很好的感光聚合物材料，然后用透明的模板压印并用紫外光照射使感光材料凝固，退模后再用离子反应刻蚀或其他方法进行图案转移的后续加工。因为不需要高温和高压，这一方法的优点十分突出。 随后，我们又提出了紫外光软膜纳米压印技术，即用软光刻的方法复制具有一定弹性和易退模的聚甲基乙氧基硅烷

（PDMS）结构作为压印软模，只需在很小的压力就可将这种软模上的图案转移到感光聚合物材料的涂层上，经紫外光固化和退模后便可得到所需的图案。与热压印相比，这种紫外光软膜纳米压印技术可更好的保证大面积纳米复制的均一性。

二、微流控技术

微流控是一种芯片技术。用微纳制造的方法可以在玻璃、塑料或硅基材料上制作与集成电路类似的微流体芯片，并用它对微量液体或气体进行精确操作并完成各种化学和生物学方面的实验。微流芯片基本结构单元的几何尺寸一般在几十到几百微米的范围之间。

虽然微流芯片可用玻璃和硅等材料制备，但用聚合物材料如塑料和PDMS制作微流芯片具有工艺简单、成本低以及生物兼容性好的优势。例如，可以将液态的PDMS及其催化剂混合物浇铸在一个用普通光刻方法制成的光胶模版上，经半小时热处理PDMS固化，剥离后可打孔，然后经等离子表面处理并粘贴在玻璃基片上即得到实用的微流芯片。在微流芯片的制作过程中，材料的表面处理具有特别重要的意义。一方面，微流通道的亲水性直接影响了微流样品的输运，通过物理或化学的表面改性可以使流体样品的浸润性提高。另一方面，表面处理的好坏，也决定了芯片键合及封装的成功与否。常用的表面处理方法如等离子清洗、化学清洗和分子自组装等方法都可以用在微流芯片的加工制作过程中来。

微流芯片可以被看作一个微量流体的操作和反应系统。最简单的芯片可由微流通道和与之相连的进样口和出样口组成。稍微复杂一点的芯片也可加入微反应池、微混合器、微分离单元、微探测器和其他的一些功能单元。功能更强一些的微流芯片有微泵和微阀，以便对微量流体进行有效的操作。但微流芯片中液体的驱动不一定是由机械泵来实现的，很多其他类型的驱动力诸如压强（液压、气动、热膨胀）、毛细力、电泳、电渗、介电力、表面声学波等都可以有效地应用于微流液体的驱动。

由于尺寸的减小，微流样品在芯片中的行为与宏观影像有着很大的差别。首先，在微流沟道中流体的速度分布是的很不均匀的。简单地说，由于黏滞力的作用，流体在通道中心的流速最大，越靠近管壁，流速越小。微流系统的尺度效应也表现在毛细效应、黏弹性、电动效应等物理过程中，从而导致一些特异的微流现象。特别重要的是，表面的作用将随着尺度的减小不断加强，这使得在宏观流动中常被忽略的效应如液体的表面张力、粒子电离后产生的库仑力、分子极化产生的范德华力、空间位形力等成为一些必须考虑的因素。如果处理的对象是比较复杂的流体，如电解质溶液、高分子溶液、有悬浮粒子的液体，甚至是凝胶液、多相流体、有化学反应的流体等，问题就会更复杂。此时，应当用宏观流体力学与微观分子动力学连接起来的介观理论来表达特征尺度为微、纳米量级的流动问题。例如用微流芯片在非溶性的两相流体中产生稳定地液体串并对其进行分裂、聚合等一系列操作。也可在微滴内进行不同的化学反应，以达到高通筛选的目的。

三、几个应用的例子

微纳制造技术的应用极其广泛。就纳米压印技术而言，它可以用来制作纳米光学器件如亚波长纳米光栅、光子晶体、半导体光源及显示屏的增光膜等。它也可用来制造超高密度磁碟、场效应晶体管、单电子器件等。纳米压印技术在生命科学中的应用主要包括基因电泳芯片和仿真细胞外基质的制作等。

组织工程和再生医学的核心是建立由细胞和生物材料构成的三维空间复合体，由此形成具有生命力的活体组织，对病损组织进行形态、结构和功能的重建并达到永久性替代。例如可用少量的组织细胞通过体外扩增，进行大块组织缺损的修复。由于可在按组织器官的缺损情况进行任意塑形，有望达到形态的完美修复。已经证明，体外细胞生长的形态受基底材料和基底材料表面纳米结构的影响。

我们注意到，现有的微纳制造工艺主要适用于合成聚合物材料如感光乳胶薄层的图案化，而理想的细胞外基质材料应该是可降解的自然生物材料如胶原蛋白（Collagen）等。传统的微纳制造方法显然对这类自然生物材料的图案化不适用。根据对生物兼容性和持续发展的考虑，我们提出了一个天然生物材料图化的新方法：首先通过负压干燥的方法将PDMS的纳米图案复制在明胶（Gelatin）薄层上，然后通过化学处理使明胶层的纳米结构固定。由此得到的自然生物材料的纳米图案可以用于细胞培养及细胞的三维支架即细胞外基质介观生物材料的仿真。

另一个例子是外周血中稀有细胞（如循环肿瘤细胞）的高效捕获。肿瘤细胞应进入血液循环就有可能发生转移，但因其数量非常稀少（每毫升病人样品或每千万个血细胞中不到一个循环肿瘤细胞），故检测具有很大的挑战性。目前所用的捕获技术有两种：一是根据循环肿瘤细胞的特定标志物来修饰磁颗粒或固体的表面，当循环肿瘤细胞与这些材料的表面发生免疫作用时就可被捕获，其他细胞则可被清洗掉。 二是根据循环肿瘤细胞与血细胞的大小或形变能力的不同来设计有一定孔径的过滤膜，当样品通过过滤膜时，循环肿瘤细胞可被捕获而血细胞则可通过。因为没有免疫作用，细胞不会受损，所捕获的循环肿瘤细胞可用来进行培养、观察及药理实验等，这种过滤膜技术被十分看好。但问题是用这种方法所截留的不仅仅是循环肿瘤细胞，且还有相当数量的血细胞，所以选择性不是太好。最近，我们用微纳制造的方法，设计并制造了一种特殊的过滤膜，使其对循环肿瘤细胞捕获的选择性大大地提高，有望在短期内进入临床试验和大规模应用。

微流控技术可用于各种生化分析和相关的物理和化学实验。微流控特别适用于细胞分析、基因测序、核糖核酸的提取和纯化、蛋白质结晶等。这一技术还提供了一种单细胞和分子操控的有效方法。无论在动物或植物体内，微流体都在无时无刻地维持或参与着整个的生命过程。可以想象，对生物体内微流体运动的介观仿真将推动生物医疗技术的深入发展。

从生物工程的角度看，微流控技术的优势一目了然。微流控芯片体积小、可集成度高、低成本和大批量生产。我们可以在同一个芯片上集成不同功能的单元，如样品提取、分子筛选、分离和分析等，从而一次性完成生物实验的全过程。也可以在同一芯片上平行完成需要重复的分析（如基因解码、聚合酶链扩增反应、蛋白质结晶等）并通过对实验参数的系统调控达到高通筛选的目的。

显然，阵列式集成的微流细胞芯片也有可能大规模地应用于干细胞分化因子的调控、生物毒性的检测及药物高通筛选等方面的试验。总之，芯片上的多通道、多功能和多系统的集成（样品处理、PCR、电泳分离、片上监测自动化与计算机化）将成为生物芯片技术总的发展趋势之一。但目前许多功能性单元需要进一步研究和优化。今后可望在高速基因测序、基因工程、蛋白质学和细胞分析等众多领域得到大规模的发展和应用。

微流控细胞分析仍然是当前的研究热点之一。我们注意到，在此类芯片中细胞培养大多是灌注的，即细胞在微流通道或与微流通道紧密相邻的微池中。由于细胞表面剪切力的限制，这种方法不利于流体的动态调节。从仿真的角度看，细胞在体内处于毛细血管之间的狭小空间中，其营养物质及信息因子的输运最终是通过扩散来完成的。因此，我们要设计类似于毛细血管的微流扩散网络以进行细胞长期培养和相应的组织结构形成。

微流控技术与纳米压印技术用很好的兼容性。将其结合，我们成功地实现了基因的快速分离。首先将不同尺寸和不同间距的纳米点阵及微流通道分别用纳米压印和光刻的技术复制到基片上，刻蚀后再将含有微流通道结构的PDMS膜与之封装。在电场作用下样品的中的基因片段会通过各个纳米点阵。由于大小不同的基因片段受到纳米列阵的阻碍是不同的，样品中的基因片段可以被很快地分离开来。而且纳米压印技术的低成本，该种纳米器件应该有一定的商业化前景。

四、小结

综上所述，微纳制造、微流控芯片及介观仿真在生命科学领域有很好的应用前景。如何利用这一机遇，率先建立交叉学科的优势和有效的应用模式，是当前必须考虑和完成的重要课题。据此，我们在认真开展各项基础研究的同时，开发了一些微纳制造和微流控芯片应用与生物医学的新设备和新方法。从所积累的经验看，这一模式是可行的，但必须更深层次地参与相关的医疗检测和临床应用。只有在基础研究和商业开发相互协调与相互促进的条件下，我们才有希望使微纳制造和微流芯片技术在生物医学领域得以健康和迅速的发展。

参考文献

［1］Y. Chen， A. Pépin. Emerging Nanolithographic Methods［M］//H.S. P. Houdy，C. Dupas and M. Lamany （Eds.）Nanoscience，Part I. Berlin：Springer 2007.

［2］A. Buguin，Y. Chen， P. Silberzan. Microfluidics：Concepts and Applications to the Life Sciences［M］//Patrick Boisseau，Philippe Houdy and Marcel Lahmani（Eds.）Nanoscience，Part 2. Berlin：Springer， 2009.

［3］石剑，陈勇．紫外光软模纳米压印技术［M］//前沿科学的交叉与融合．北京：北京大学出版社，2008.

［4］陈勇．微流芯片技术与微流芯片实验室［M］//前沿科学的交叉与融合．北京：北京大学出版社，2008.

陈勇简介

1977年考入武汉大学物理系半导体专业学习。1986年11月获蒙彼利埃大学博士学位。1990年任职法国国家科研中心主任研究员，1998年为二级主任研究员，2010年一级主任研究员，2003年到2008年间年为巴黎高等师范教授，还一级主任研究员，先后任中科院海外评审专家，北京大学长江讲座教授、日本京都大学特聘教授（PI）、香港理工大学讲席教授、江汉大学特聘教授及一些高校及科研院所的客座教授或科技顾问；参与创建巴黎及武汉介观生物等高科技公司。发表科技论文500多篇，参与多个欧盟框架及其它科研项目并指导过40多位博士论文的研究。

汽车发动机的发展与展望

汪善进

雷诺汽车公司发动机研发部

1 Allée Cornuel 91510 Lardy, France

当今世界开汽车的人多，懂汽车的人少，懂汽车发动机的人更少。本文介绍一些汽车内燃发动机的基本常识和重要技术，对目前热门的新能源汽车发动机做了比较全面的介绍和分析，并对汽车发动机的发展方向，特别是对中国汽车发动机发展现状和前景提出了一些独特的见解。

1. 内燃发动机简介

在汽车被发明之前，人们用的是马车，因此汽车可以说是“无马之车”，而代替马的正是汽车发动机。法国可以说是世界上第一辆汽车的诞生地，因为早在1769年，法国人居纽（F.J.Cugnot）就制造了世界上第一辆蒸汽驱动三轮汽车。但人们一般都公认德国人苯茨和戴姆勒为汽车工业的鼻祖，因为他们于1886年制成了世界上第一辆真正可以长时间高速行驶的三轮汽车。与法国人的蒸汽驱动三轮汽车不同，德国人装的是他们制造的汽油内燃发动机。而这台发动机的功率仅为1.1马力，车速只能达到18千米/小时。

有了内燃发动机，才有了汽车。汽车的发展史实际上就是内燃发动机的发展史。简单讲发动机就是一个能量转换机构，即将汽油（柴油）或天然气的热能，通过在密封气缸内燃烧气体膨胀时，推动活塞做功，转变为机械能，驱使汽车行驶，这是发动机最基本原理。这种能量的转化主要是靠活塞的往复运动完成的，所以叫作往复活塞式内燃发动机。

内燃发动机是由气缸体、气缸盖、活塞、连杆、曲轴、飞轮等主要机件组成的。活塞安装在圆柱形的气缸内，可在气缸内上下移动。连杆一端与活塞连接，另一端安装在曲轴的连杆轴颈上，曲轴由气缸体上的轴承支承，可在轴承内转动。当活塞受高压气体作用，由上向下移动时，通过连杆推动曲轴旋转。就这样，直线运动被巧妙的变成了旋转运动。

内燃发动机的一大天生的缺陷：热效率低。热效率低费油不说，发动机的动力（或功率）也很难加大。按照热力学原理，将热能转化为机械动力的热能机的效率总是有限的。理论上说，一台四冲程内燃发动机的效率不可能超过55%，而现实中的内燃发动机，因为有摩擦和热量散失，效率一般在30%左右，也就是说有将近3/4的能量被浪费掉了。

2. 内燃发动机的主要技术

一台内燃发动机有将近1000个部件，重要部件也有几百个。每个部件都是技术的结晶，都是汽车工程师几年甚至几十年的研究和试验成果，发动机里所采用的技术可以说不计其数。即使把每个部件做好了，也不等于造出的发动机性能就好，因为还需要进行整体优化。对现代汽车来说，光有好的发动机硬件还不行，还需要一整套发动机软件来控制，以保证发动机能在最优化的状态运行。这也是高端的技术。

发动机的性能指标有多种：可靠性和耐久性指标、动力性指标（有效转矩、有效功率）、经济性指标（有效燃油消耗率）、环境指标（废气排放和噪声水平）、舒适度指标（平稳运行和振动水平）。自从汽车被发明以后，人们首先要解决的是发动机的可靠性和耐久性，要知道刚开始的汽车发动机开不了几千米就会抛锚，现在的发动机有的能完好地开到一百万千米！如何增加发动机的动力、减少发动机的震动和噪音、降低油耗以及减少废气排放，都是汽车工程师一直在努力解决的问题。这一切都需要技术。下面介绍的是目前还在开发的影响较大的几项技术。

2.1 强制增压技术

强制增压的作用是增加发动机的动力但不增加发动机大小。强制增压发动机强制空气以比平常更高的压力，把更多的空气推进燃烧室，从而获得更大的功率。因为体积小的发动机摩擦小、损耗小，因此也更节省燃油。同样大小的发动机，采用强制增压技术，功率可以增加50%，甚至加倍。强制增压有多种方法，用得最多的是涡轮增压器。但涡轮增压器在发动机低速区域运行不佳，在发动机高速运转时才能起到大的作用，这对在低速区域扭矩较大的柴油发动机来说特别合适。涡轮增压器从20世纪60年代在汽车发动机上得到应用，目前100%的柴油汽车发动机使用这项技术，在汽油发动机里也用得越来越多。

2.2 可变气门定时技术（VVT）

VVT指的是发动机气门升程和配气相位定时可以根据发动机工况做实时的调节。主要作用是提高汽油发动机的低速扭矩，增加低速区的动力，并使发动机在中小负荷时有优异的热效率，达到节省燃油目的。

2.3 汽油机直喷（GDI）和稀薄燃烧技术

这项技术的主要作用是节省燃油。将高压汽油直接喷射到气缸内，周围为稀薄混合气，实现分层燃烧，可提高燃料经济性，节油能达到20%。

2.4 停缸技术

停缸技术道理很简单，就是当发动机在中小负荷运行时，使一部分气缸停止工作，可让其他气缸的工作效率提高，从而达到节油目的。

2.5 柴油发动机技术

从节油指标看，柴油发动机可以说是一项最有效的技术。一般柴油机的油耗要比汽油机的低30%～40%，迄今为止还没有任何其他一项技术这么有效。原因主要有二：首先，每单位柴油的能量含量比汽油高；其次，柴油机的压燃特性，使其热效率比汽油机高。柴油机还有另一大优点：低速时扭矩大，为汽车提供了更好的使用性能，同等马力的柴油机，在低速区域开起来感觉要比汽油机力气大得多。这就是为什么在油价昂贵的欧洲，买柴油机汽车的要比买汽油机汽车的人多。

柴油机最大的问题污染和噪声。机动车的主要排放物有一氧化碳、碳氢化合物、二氧化碳、颗粒物和氮氧化物。相对而言，柴油机的一氧化碳、碳氢化合物和二氧化碳排放量极低，但颗粒物和氮氧化物的排放要大得多。处理颗粒物和氮氧化物要比处理一氧化碳、碳氢化合物难得多，需要更尖端的复杂的技术，成本很贵。目前主要采用加装氧化型催化转化器和研究开发NOX催化转化器以及具有良好再生能力的微粒捕集器。随着欧洲和世界其他各地对汽车废气排放标准的要求越来越严，柴油机后处理的代价会越来越高。

与汽油机的点火燃烧不一样，柴油机的燃烧是靠加压到一定程度柴油自爆自燃的，因此产生的噪音要大得多。若不采取特殊措施，同等马力的柴油机发出的噪音要比汽油机高出10dB甚至20dB（高10dB相当噪音增加8倍）。如何减小柴油发动机机的噪音和振动曾经是汽车工程师们的一个多年的难题。自从20世纪80年代发明了柴油高压喷射和电控喷射技术后，柴油机的噪音才得到了有效的控制。控制柴油机的噪音除了需要高精度的高压电控喷射系统外，还需要一整套柴油喷射优化方法和软件。到目前为止，还有不少柴油发动机发出很大噪音，让许多亚洲和北美洲的顾客无法接受，但做得好的柴油机（比如说雷诺新款的1.6 L柴油发动机），其噪音和振动水平已与汽油机不相上下。

3. 新能源汽车发动机

目前，全世界都在担心大气变暖，汽车的CO_2排放越来越成为众矢之的。另外，石油的储存短缺也是也决定了内燃机汽车不可能永远统治汽车市场。怎样用新能源来取代石油原料，成了汽车制造商在短期内必须解决的问题。目前的新能源汽车发动机主要有以下几种。

3.1 纯电力驱动技术

纯电动汽车是指以车载电源为动力，用电机驱动车轮行驶。一般采用高效率充电电池为动力源。电动机的发明实际上比内燃机还要早，因此电动汽车的历史比内燃机汽车还更长。电动汽车之所以至今还没被普及是因为其有两大缺点：汽车的连续行驶里程太短，价格太贵。

电动车不耗油，但必须充电，而充电要有蓄电池。蓄电池技术有很多种，目前单位储存能量最高的蓄电池是锂离子电池。1千克锂离子电池可以储存80~200瓦时的电。电池重量越大，储电量越多，汽车的连续行驶里程就越长。但若电池重量太太，不仅成本太高，而且会过分增加汽车重量和体积，使得汽车行驶起来更加费电，反而会减少汽车的续驶里程！对现代汽车来说，最佳的锂离子电池重量在100~200kg之间。用这样的电池，汽车的续驶里程只能达到150km左右。法国的雷诺公司是发展大众型纯电动车的先锋，目前有4款大小不同的纯电动车同时在欧洲市场销售，设计新颖，性能良好，价位已到达大众能接受的水平，但续驶里程都在180km以下。按专家估计，若续驶里程能达到500km以上，普通老百姓才会愿意买电动车，但目前的电动车离这个目标还相差太远。如果不能发明出重量更轻、能量更高的蓄电池技术，这个目标就不可能实现。

电动汽车也有很多优点。电动机的扭矩平稳、转数范围很大，而且噪音很小，因此电动车开起来给人的感觉是连续、平稳、安静，舒适度要比内燃机汽车高。另外，前面说过，内燃发动机的效率很低，将近3/4的能量以热能的形式散发掉了，而电动机的效率能到达90%，能量浪费较少。纯电动汽车最大的优点是它的CO_2排放等于零，因为油消耗量是零。当然，这并不意味着使用电动车完全没有CO_2排放，因为发电也会产生CO_2，特别是火力发电。法国的电80%来自核能，按这个比例计算，一辆纯电动车的CO_2排放量大约是30 g/km，远远低于一辆同样大小的内燃机汽车的排放量。而在中国，因为发电主要靠烧煤，一辆纯电动车的CO_2排放量能到达130 g/km。即使如此，这个数字仍然比同样大小的内燃机汽车的CO_2排放量小（在欧洲，一辆内燃机汽车的平均CO_2排放量高于130 g/km，这还没算石油的生产和运输所排放的CO_2）。此外，开纯电动车确实可以解决城市污染问题。

电动车需要充电，而充电需要发电。若靠烧煤来发电的活，显然得不偿失，不可持续发展。法国的电主要来自核能，核电站生产的电白天供应给工厂，晚上用来给汽车充电，是一个可持续的模型。但无论以如何方式发电，电量总是有限的，这就注定了同时充电的汽车数量也是有限的。因此，纯电动汽车的发展尽管是大势所趋，但主导汽车市场的可能性不大。

3.2 混合动力技术

混合动力车采用传统的内燃机加上电动机作为动力能源，通过混合使用热能和电能两套系统开动汽车。混合动力系统的最大特点是油、电发动机的互补工作模式。在起步或低速行驶时，车子仅依靠电力驱动，此时汽油发动机关闭，车辆的燃油消耗量是零；当车辆行驶速度升高（一般达40 km/h以上）或者需要紧急加速时，汽油发动机和电动机同时启动并开始输出动力；在车辆减速行驶时，混合动力系统能将动能转化为电能，并储存在蓄电池中以备下次低速行驶时使用。

混合动力车的主要优点是省油。一辆使用汽油的混合动力车的油耗要比一辆同样大小的汽油机汽车低30%～40%。从CO_2排放量来看，一辆柴油机汽车与一辆使用汽油的混合动力车不相上下。因此，如果使用柴油机与电动机的混合动力，CO_2排放量还可以减少好多。但柴油机的噪音大，要与无噪音的电动机搭配好难度较大，生产成本也更高。目前除标致雪铁龙外，绝大部分混合动力车只用汽油。

混合动力车的主要缺点是成本高，因为至少有两台发动机（有的车装有3台发动机），再加一个小的蓄电池，造价要比一般的内燃机汽车高5000～10000欧元。成本昂贵、省油有限，这也注定混合动力车不可能在短期内占据太大的市场。

3.3 氢燃料电池驱动技术

氢燃料电池驱动技术可能最具有取代内燃发动机的潜力。氢燃料电池通过氢和氧的化学反应产生电能，这种化学反应的唯一副产品是热量和水蒸气。电能供应给电机来驱动汽车。相对于内燃机驱动，燃料电池驱动的效率更高，污染更低，甚至是没有污染，它排出的仅有纯净的水蒸气。实际上从20世纪60年代起，航天工业就已经使用这种燃料电池了。该技术已应用于部分试验车辆和为某些建筑物提供电能。但由于氢燃料电池成本很高，而且氢燃料在液体储存时需要很高的压力（700 bar），有安全问题隐患，目前商业化推广还为时过早。即使哪天氢燃料电池动力能成功地应用于汽车，还有一大难关要过：那就是如何大量生产燃料电池所需要的氢？

4. 汽车发动机的发展前景和挑战

汽车的持续发展面临着许多问题和挑战，目前公认的最大的问题是大气变暖和CO_2排放。如何找到一条既有效又经济的办法降低燃油消耗量和CO_2排放，将成为汽车工程师要解决的主要问题。目前，混合动力技术的节油程度有限，电动汽车需要更新的蓄电池技术，氢燃料电池驱动技术尚不成熟，没有哪一项技术能够取代传统的内燃机。因此，将来的市场尽管会有越来越多的新能源汽车，但内燃机在未来的30～40年仍然会占据主导地位。

为了强迫汽车制造商采取有力的减排措施，欧盟制定了世界上最严的CO_2排放标准。按照这个标准，到2020年，每家汽车制造商必须保证它在欧盟销售的小轿车的平均CO_2排放不能超过95 g/km，否则要交一大笔罚款。为了到达这个标准，汽车制造商除了对传统的内燃机汽车进行大量改进之外，还必须配上一定数量的电动汽车和混合动力车。

传统的内燃机经过100多年的发展，技术已非常成熟。即使如此，汽车工程师还是有办法使它的CO_2排放有明显的减少。除了采用以上所介绍的几项节油技术外，主要办法是靠优化：优化发动机的运动部件设计以减少摩擦，优化燃烧和热量控制系统以减少热量损耗，优化发动机运行控制系统以便让发动机在最省油区域运行，等等。还有，减轻汽车和发动机的重量也是一大发展方向。通过这些优化技术，内燃机的CO_2排放有望再减少20%到30%。在欧洲，尽管有些环保和卫生组织极力反对柴油车，因为在节油方面有巨大优势，柴油机汽车在还将会继续占领较大的市场份额。

至于电动汽车，最大的难题是怎样大大地提高它的续驶里程。在新的蓄电池技术被发明之前，一种可行的办法是在纯电动汽车的基础上再加一小型的内燃机。这个内燃机的功能不是驱动汽车，而是在蓄电池的电用完时负责给它充电，这样可以把电动汽车的续驶里程延长到400km以上。如果汽车短距离行驶，内燃机不起作用，与纯电动汽车一样，没有CO_2排放。这项技术的优点很明显，解决了纯电动汽车续驶里程低的缺陷，让顾客需要长距离行驶时可以放心地开车，不会担心半路上没电。缺点是造价高，因为要加一台发动机。鉴于这项技术的优势，它的发展前景应该是很好的。

再来说混合动力汽车，它的主要缺点是除了造价高外，对减少CO_2排放作用有限，因为汽车行驶时大部分时间还是靠内燃机驱动。要进一步减少CO_2排放，必须得让电动机工作时间加长。有一种办法是在混合动力汽车的基础上加一可充电的蓄电池，蓄电池充满电后，可以让电动机在更大的范围内取代内燃机驱动，从而进一步减少CO_2排放，这就是所谓的可充电的混合动力汽车。

5. 中国汽车发动机发展现状和展望

2013年中国的乘用车产销接近1800万辆，稳稳地排在世界第一。但在这个显赫的数字背后，隐藏着中国汽车工业发展的巨大忧患。在小轿车市场，外国品牌占据统治地位，国产品牌远远被抛在后面，短期内赶上国外品牌的希望渺茫。至于汽车发动机，国产品牌与国外的差距更大，尽管目前国内已经能够自行设计并开发中小功率的汽油发动机，但好一点的车仍然依靠引进外国的发动机技术或购买国外的发动机。为什么是这种状况？根本原因是三个字：技术低。为什么我国的汽车发动机技术低？笔者认为原因主要有三：第一，发动机是一项非常复杂的技术，不是一两年就能赶得上的。第二，国外的汽车制造商非常注意保护发动机的关键技术，不会轻易转让。第三，我国至今还没有足够重视发展汽车发动机技术。前几年，中国政府的汽车发展策略是：重点投资开发电动车，因为电动车我们与西方差不多站在同一条起跑线上，而传统的内燃机，我们起码比西方晚30年，短期内很难赶上，干脆不赶了。这种策略听起来有道理，但会带来很大的问题。首先，纯电动车在可预见的将来只能占据小量的市场份额，若搞纯电动车而放弃内燃机汽车，等于捡了芝麻丢了西瓜；其次，如果哪天中国选择发展混合动力技术，那也得首先掌握好内燃机技术。内燃发动机技术上不去，我们这个世界第一汽车大国将会继续处在技术弱国地位，继续受外国品牌统治。

因此，大力发展本土的汽车发动机工业，尽快全面掌握内燃发动机技术，赶上国际先进水平，这是中国汽车发展的当务之急。内燃机是最基本也是最复杂的技术。欧美的汽车发展有100多年的历史，日本人起步晚，但也有80多年的历史。可想而知，要想达到他们的技术水平，不是几年之内就能成功的。现在国内有的汽车公司试图通过收购国外品牌来获取关键技术（比如收购沃尔沃），也有的与国外的发动机开发公司合作开发新款发动机（比如奇瑞和奥地利AVL公司的合作），这些都是捷径。但笔者认为，要想真正让中国的汽车企业进入世界列强行列，光靠购买别人的技术是不行的，光靠一两家民营企业单打独斗也是不行的。首先国家需要制定宏大的长远的汽车发展战略，采取强有力的措施使国内的企业有动力去吸收国外先进技术并大力自主开发汽车发动机；其次是国内的汽车企业要整合重组，集中人力物力形成几家大规模的企业，才有与当今国际汽车列强竞争的可能性；最后也是最重要的，是要培养一大批发动机和汽车的专业人才，短期内可以采取政策吸收有经验的海外人才。有了人才，才能真正掌握技术，才能去创新，才能达到世界先进水平。韩国的汽车发展很值得我们借鉴。20年前，韩国的汽车和发动机技术状况与现在的中国差不多，但20年后的今天，韩国的现代汽车集团已经进入世界的前5强，韩国已是名副其实的汽车强国了。韩国人能做到的，中国人也应该能做得到！

作者简介

汪善进，1963年4月出生于江西。1979年考入清华大学机械系，1984年大学毕业。1985年赴法国留学，就读于巴黎中央理工学院，1990年获该校材料科学博士学位。1990年至2000年在法国GKN汽车传动轴公司工作，先后担任科研工程师和课题组长。2000年开始进入法国雷诺汽车公司，从事发动机开发和研究，主攻发动机的振动和噪音（NVH），先后担任研究组长、发动机NVH专家。现任雷诺公司发动机研发部NVH专家、NVH技术总负责人。

作为雷诺公司发动机NVH部的技术领头人，带领团队制定了所有发动机主要部件的NVH设计规则；参与了所有雷诺新型发动机和变速箱的NVH设计和改良，包括新一代电动发动机（雷诺公司的中小型柴油发动机被专家认为是欧洲最好的发动机，很大一部分归功于其优良的NVH性能）；拥有多项发动机技术发明，并获雷诺公司技术发明奖。

曾多次被邀请在国际汽车噪音与振动大会上做专家发言和重点报告，现为奥地利国际汽车噪音与振动大会（ISNVH）科学委员会成员。2009年参与组建了清华大学法国校友会，并担任校友会会长至今。

抑郁症患者互助行为调查结果分析[①]

王思萌

巴黎高等师范学院，社会科学学院，博士生候选人

Département de Sciences Sociales, École Normale Supérieure, Paris, France

摘　要：目的：了解抑郁症患者的互助情况。方法：于2010年1～3月在北京一家精神病医院选取了经过诊断的抑郁症患者，共177人填写匿名调查问卷；之后对其中15人进行了深入访谈。结果：49.7%的调查对象有帮助过病友的行为，其中帮助过1～3名病友的比例占68.7%；52%的调查对象被病友帮助过，其中被1～3人帮助过的调查对象占75.6%；属于既帮助过其他病友又接收过其他病友帮助的调查对象占22.6%；参加过固定互助组织的调查对象占24.3%；涉及患病经历和治疗经历内容的谈心和聊天是调查对象参与最多的互助活动形式；绝大多数调查对象对病友互助持有积极的认可态度，认为此类互助既有助于自己对他人倾诉心理困扰，又可以交换涉及治疗和其他应对抑郁症的有用信息。结论：病友互助属于内在社会支持范畴，与来自政府机构、医疗机构、家庭，亲戚、社区、友邻或同事的外在社会支持相比，患者群体中内在的社会支持有其独特的益处，体现了患者应对疾病的主动性、自主性及建构互惠网络的能力。

关键词：抑郁症　病友互助　社会支持

---

① 本研究项目已通过该精神病医院伦理审查委员会的伦理审查，并全程接受其监督。问卷调查与深入访谈经过研究对象的知情同意。

世界卫生组织最新调查统计分析，全球抑郁症患者已达3.4亿，在年满20岁的成年人口中，抑郁症患者正以每年11.3%的速率增加。在我国，与心理应激相关的患者约占全人口的5%～10%，已有2600多万人患抑郁症。但是，我国每万人口仅有精神科床位1张，每10万人口约有精神科医师1人，与全球平均水平相比，仅为四分之一。这种严重的供需不平衡成为我国亟待解决的公共健康问题之一。在这种情况下，考察抑郁症病友互助的模式，对于缓解此类精神病患者的病情具有一定意义。自从美国精神科与社区医学教授Sydney Cobb[1]在1976年的一次演说中系统讨论了社会支持对人类身心健康的影响之后，各种衡量社会支持和健康关系的量表和问卷层出不穷，到目前已经成为公共卫生科学界广泛认可的研究领域。

在我国，涉及患者的社会支持研究大多针对了来自患者群体外部的社会支持（肖水源等[2]，1987）。当一个人遭遇心理危机时，能否将其化解取决于以下三方面因素："社会支持网络的规模、密度和异质性程度；社会支持力量的强度；被支持者的个性特点"（李强[3]，1998）。抑郁症社会支持的最大来源仍然为配偶及其他家庭成员（管新丽[4]，2003；田旭升等[5]，2006）。涉及以病友为载体的社会支持之研究尚不多见，属于一个亟待开发的研究领域。少量直接针对来自患者内部支持的研究尚未针对精神病患者，而是在喉癌切除病友（郑淑君等[6]，2004）、乳腺癌病友（裘佳佳等[7]，2008）、艾滋病患者中展开。病友互助小组中患者能够相互交流、相互心理扶持，积极参与治疗和护理，对患者心理康复和逐步回归社会有积极作用。

一、对象和方法

这尚在深入的研究属于一个医学社会学和精神卫生研究相结合的范畴之内，主要研究方向包括两个内容：一是抑郁症患者的互助行为，二是患者互助的组织机制。在本文，我们将集中讨论抑郁症患者之间的互助行为。本课题的研究对象均为北京某精神病医院门诊和病房的抑郁症患者。所有研究对象已经被正式诊断为抑郁症患者。在2010年1月～3月期间，我们共在这家医院发放177份匿名问卷。我们使用的《病友互助调查问卷》经过预调查和多次修改后定稿。该问卷分三部分，第一部分为一般信息，包括性别、年龄、教育程度、婚姻状况、职业。第二部分涉及病友互助行为及参与互助组织的情况，具体问题包括“是否帮助过病友”“是否被病友帮助过”“是否参与过互助组织”“互助组织的规模的有多大”以及互助组织成立的时间和活动频度等。第三部分涉及研究对象对病友互助的认知及态度，包括研究对象对病友互助内容和形式的了解，对病友互助之益处的认识以及是否赞同并支持病友间的互助。

在收回问卷之后，我们使用SPSS软件对收集到的数据予以了统计学处理。我们应该指出，本研究尚处于一个初始和尝试阶段，所以我们没有使用典型的抽样法，而是在选择调查对象的过程中采取了志愿参与既吸收的原则。但是我们所接触到的患者（其中包括一部分没有过互助经历的患者）大多非常踊跃地表示愿意参加该研究并最终积极地填写了问卷。所以，我们认为，填写问卷的患者样本和该院门诊部、住院部的患者应该具有相当的同质性。

二、结果

在我们的研究对象中，男女比例为4：6；处于19岁～28岁、29岁～38岁、49岁～58岁年龄组的人合计占到了84.2%。至少上过大学的人占54.9%；未婚者、离婚者、丧偶者加起来的比例为44.1%；学生占15.8%，无业人员占14.9%，退休人员占10.2%，农民仅占0.6%，其余的人均有不同类别的工作（表1）。从上述信息中我们可以看到，我们的研究对象属于一个受教育程度较高的患者群体。但是，我们同时可以看到，无业和退休人员的比例也很大，共占1/4。

表1　受访人的人口学和社会背景资料

年龄	18岁及以下5.7%	19岁～28岁26.0%	29岁～38岁20.3%	39～48岁17.0%	49～58岁20.9%	59岁及以上10.2%
教育程度	小学3.3%	初中11.3%	高中30.5%	大学45.8%	硕士8.5%	博士及以上0.6%
婚姻状况	未婚34.5%	已婚54.2%	离婚6.8%	再婚1.7%	丧偶2.8%	
职业	无业14.9%	学生15.8%	教师9.0%	医护人员4.5%	会计3.4%	工程师4.0%
	公务员4.5%	企管人员5.1%	普通职工19.2%	自由职业7.9%	退休10.2%	农民0.6%

我们在上面所讲到的互助行为发生方式包括四种：第一是个人行为，以一人对一人为主；第二是通过自发的QQ群或其他网络方式参与有组织的互助活动；第三是参与医院组织的互助活动；第四是参与患者在现实生活中自发组织的互助活动。我们发现，接近半数的患者帮助过病友，略超过半数的患者被病友帮助过，接近三成的患者以上两种行为皆有，即具有严格意义上的互助行为。更为细致地讲，49.7%的调查对象帮助过其他病友，其中帮助过1～3名病友的比例占68.7%。同时，52%的调查对象表示曾被其他病友帮助过，其中被1～3人帮助过的调查对象占75.6%；属于既帮助过其他病友又接受过帮助的调查对象共占22.6%。

在我们的研究对象中，参加过固定互助组织的人占24.3%，主要包括基于虚拟空间的组织和基于真实空间的组织两类，规模从几个人的小型互助组织到上千人的网络组织不等，参与时间从1个月至3年不等，活动次数从1次至1天1次不等，活动形式多为见面聊天或上线聊天。在有组织的互助活动中，最常见的人数规模是20人左右，其中参与时间半年至一年的患者最多，共占33.3%。从研究对象提供的病友互助内容来看，交流治疗经验和提供心理扶助占主导地位，但提供治疗信息和情感支持也是很重要的内容。病友之间的互助还包括了物质支持，虽然这种行为发生的频率并不高（表2）。

表2　病友互助所包含的内容

互助内容	回答		样本百分比
内容分类	回答人次	百分比	
治疗经验	139	32.0%	79.9%
治疗信息	89	20.5%	51.1%
心理扶助	108	24.9%	62.1%
情感支持	83	19.1%	47.7%
物质支持	14	3.2%	8.0%
其他内容	1	2%	.6%
合计	434	100.0%	249.4%

在态度方面，64.4%的患者对病友互助抱有支持态度，10.2%持否定态度，其他患者持观望和有待进一步了解的态度。在对病友互助的益处所在问题上，我们的调查对象心中显然有一个轻重地位的排序。第一位是有治疗相关的信息和经验交流（赞同率42.5%）；第二位是倾诉心理困扰（赞同率40.8%）；第三位是通过互助结交朋友（赞同率10.9%）。这就是说支持患者互助的研究对象，认为它的最大益处在于交换康复信息，同时倾诉各自的心理困扰。

三、讨论

目前，无论国内还是国际上对与患者有关的社会支持之研究多从政府、医院、家庭、亲属、社区、朋友、邻居，或媒体环境等角度加以考察，而常常忽略产生于患者群体内部的“内在社会支持”。我们认为，病友之间拥有共同的疾病体验作为彼此沟通、交流、劝导的条件，具有其他社会支持不可替代的功能。到目前为止，我们的研究重点还放在抑郁症患者的互助行为和互助组织两个问题上。当我们认为在此基础之上，今后的研究至少可以有两个侧重点：一是对患者互助行为的量化研究；二是以队列方法探索患者互助是否能够提高治疗效果。

参考文献

［1］Cobb Sydney. Presidential Address，1976：Social Support as a Moderator of Life Stress［J］. *Psychosomatic Medecine*，1976，38（5）：300–314.

［2］肖水源，杨德森. 社会支持对身心健康的影响［J］. 中国心理卫生杂志，1987，4：183–187.

［3］李强. 社会支持与个体心理健康［J］. 天津社会科学，1998，1：92–94.

［4］管新丽. 抑郁症病人的社会支持状况及护理干预［J］. 护理研究，2003，8：902–903.

［5］田旭升，程伟. 医学社会学视野下的抑郁症变奏［J］. 医学与哲学（人文社会医学版），2006，7：26–27.

［6］郑淑君，林芳宇，李思勤，等. 建立喉癌全喉切除术病友互助小组的作用探讨［J］. 解放军护理杂志，2004，3：58–59.

［7］裘佳佳，胡雁，黄嘉玲，等. 乳腺癌康复互助志愿者病友支持方式的应用及效果［J］. 中华护理杂志，2008，8：690–693.

作者简介

王思萌，2010年毕业于清华大学社会学系，同年进入法国巴黎高等师范学院（École Normale Supérieure）攻读社会学博士学位。研究主题为“在巴黎中国移民的精神健康问题”，这一选题建立在其硕士论文基础之上（参见《社会学视野下的抑郁症患者互助研究》，王思萌，清华大学，2010）。

学术研究方面，目前已在法国、加拿大社会科学类核心杂志上发表学术文章（法语撰写）四篇，另在各类非学术类杂志上（法语撰写）发表书评、时评五篇，同时担任巴黎政治学院出版的《中国研究》杂志特约撰稿人。自2011年起，连续三年面向法国高等社会科学院（EHESS）健康社会学硕士生教授《跨文化精神病学要素》课程。2018年在巴黎高等师范学院新加开《关于中国社会的社会科学研究综述》课程（面向本硕博），以及在巴黎十三大的《移民与健康》课程（面向本科生）。此外，还积极参与了各类学术论坛及研讨会（如在法兰西学院举办的“做中国研究的青年学者社会科学研讨会”等），共计发表演讲24次，就博士论文研究的不同成果进行了介绍。另外，作为组织者和同事共同举办了三次学术论坛，主题涉及“针对温州移民后裔的青少年精神健康治疗”“在法各国移民的健康状况”等。

社会生活方面，自2011年底担任清华大学法国校友会副会长一职，主要负责学术交流和公关事务。2009年荣获巴黎使馆教育处的高级访问学者奖学金，受巴黎政治学院邀请进行了“中法社会抑郁症治疗的对比研究”。2011年7月荣获法国“健康与社会”年轻学者奖学金，受北京大学第六医院邀请进行了“中法精神健康领域机构的对比研究”。2012年荣获法国优秀法国留学生奖学金。自2012年10月起，多次就中国社会及移民问题受法国本地媒体采访或报道（《世界报》*Le Monde*，《新观察家》*Le Nouvel Observateur*），并在*France Culture*、*Vivre FM*等电台上接受采访。

人工视觉假体研究与发展概况

杨　元

巴黎高科高等电信学院，巴黎

摘　要： 视觉是人类认知世界的重要手段，人类获得外部信息中视觉信息占80%以上。为了能使失明患者重获光明，国际上不少研究小组目前都在致力于研究人工视觉假体。本文通过对人工视觉假体研究和发展的回顾，重点介绍了常见的几类视觉假体和它们存在的优缺点。

关键词： 人工视觉假体　视觉生理　视皮层　视网膜　视神经

1. 引言

视觉是人类认知世界的重要手段，有研究表明通过视觉获得的信息占人类获得外部信息的80%以上。然而，世界上很多不幸的人正遭受着不同程度的视觉障碍带来的痛苦。中国是世界盲人最多的国家。据统计，每年中国新增45万盲人，这意味着几乎每分钟会出现一例新的盲人。因此，如何实现视觉功能的修复，使盲人重见光明，就成了一个从20世纪50年代开始，世界上就有科学家开始致力于视觉功能修复的理论和临床研究。特别是随着医学与工程学的结合发展，研究人员试图通过工程的手段研制视觉假体，让盲人重见光明。

2. 几种常见视觉假体的国内外研究概况

视觉假体是一种可将图像信息处理、编码，通过微电极阵列对视觉神经系统进行作用，从而在视觉中枢产生光幻觉，恢复盲人视力的人造器官。根据视觉产生的机理，视觉假体的实现可通过对视觉通路上任意功能完好的位置进行神经电刺激，使盲人产生视觉感受。目前国际上，按照视觉假体电极的植入位置将视觉假体可以分为视皮层假体、视神经假体和视网膜假体三大类。其中，视网膜假体按其在视网膜的位置又可以分为视网膜上假体和视网膜下假体。

2.1 视皮层假体

视皮层假体的微电极阵列是直接植入在大脑的是视觉皮层上的，通过刺激视觉皮层使失明患者产生光感。对于这一技术的研究最早可以追溯到1929年代：德国神经外科医生Otfrid Foerster通过刺激清醒人大脑皮质使其感受到了光感[1]。20世纪60年代科学家正式开始了皮层视觉假体的研究。1968年，Brindley和Lewin最早进行了长期刺激视觉皮层的研究。他们通过将80个表面电极长期植入到严重失明患者的视皮层表面，发现电刺激视皮层可以引发不连续的光感[2]。后来，Srivastava 等人通过对刺激皮层的位置和患者感受到光幻点视野分布的研究得到了一系列重要的发现：刺激区域与幻点视野分布的对应关系。

2.2 视神经假体

视神经假体是在功能保持完整的视觉通路近端植入刺激器，即在有功能的视网膜神经节细胞轴突的延伸部分 —— 视神经进行刺激。这种方法避免了对大脑造成的侵入性破坏，而且是一种更简便更容易操作的技术。在中国，由任秋实、柴新禹等著名专家主持的“C—Sight计划”（中国视觉工程）首次提出了一种刺入式多电极陈列视神经视觉假体，它以特定结构的刺入式多电极阵列插入视神经作为神经接口，将被编码的电刺激耦合进神经节细胞的轴突，用于视力恢复[3]。C—Sight研究小组通过在兔子、猫等动物模型上进行了一系列视觉电生理实验，现已证明这种高密度刺入式微电极阵列既能降低电流对视神经的损伤，也能相应提高视觉空间分辨率，极大地推动了人工视觉假体研究在中国的发展。

2.3 视网膜假体

视网膜刺激器是在视网膜下或视网膜表面植入不同微电极，使外界光线转换成电流，通过微电极刺激并激活视网膜神经细胞及其网络，而产生光感。这种装置可使失明或接近失明的眼重新获得部分有用视力。视网膜假体根据电极。神经元接口解剖学位置的不同，基本可分为视网膜上假体和视网膜下假体。

2.3.1 视网膜下视觉假体（Subretinal prosthesis）

视网膜下植入法主要由美国芝加哥的Chow研究小组和德国Tubingen的Zrenner小组等研究。Chow研究小组已植入到10名患者并进行短期观察。视网膜下视觉假体是将微光电二极管阵列或半导体微光电二极管阵列置于视网膜后巩膜与双极细胞之间，该电极阵列将进入眼球的入射光转化为分级电位从而刺激双极细胞产生光感。假体处于双极细胞层和视网膜色素上皮细胞之间，假体芯片充当视网膜中光感受器的角色，光信号经自然传送途径到达芯片，在光电二极管中产生电流，即用光电转换的方式激活微电极，从而对双极细胞产生刺激。这种视网膜假体用微电极代替了受损伤的感光细胞，利用了仍然具有功能的视网膜神经网络，使用视网膜自然的编码过程。

2.3.2 视网膜上视觉假体（Epiretinal prothesis）

视网膜上假体植于视网膜神经节细胞（RGC）的顶部，直接刺激神经节细胞轴突，不需要视网膜感光的神经网络，只要求视神经和视网膜内层功能正常；而对大多数患者，即使患有视网膜色素变性多年，但神经节细胞和视网膜内的其他细胞仍然具有功能。另外，获取图像信息的CCD摄像头相当于替代了丧失功能的眼部光学系统，白内障患者适合于实施此类手术。这种假体还可以利用玻璃体腔内的液体对植入刺激器进行散热，且外置的信息采集装置更便于外部控制。相关的研究组有哈佛大学与麻省理工学院共建的研究中心；美国南加州大学、北卡罗来纳州立大学和美国橡树岭国家实验室（ORNL）、阿尔贡国家实验室（ANL）、劳伦斯利弗莫尔国家实验室（LLNL）、洛斯阿拉莫斯国家实验室（LANL）、SANDIA国家实验室（SNL）等5个国家重点实验室联合，由Humayun领导的小组；还有德国的北部德国联合研究体（Northern German Consortium）；日本Nagoya大学的Yagi等。由Humayun领导的小组从2002年开始至2004年先后成功地将4×4电极阵列长期植入6位色素性视网膜炎盲人。临床的结果报告单个电极可以诱发出可靠的光幻点，且被植入者可以分辨简单的形状并识别一些简单的刺激。

3. 各种视觉假体的优缺点

3.1 视皮层假体优缺点

视皮层假体的优点在于它是恢复视网膜和视神经损伤的患者视觉的唯一方法。当时作为一种颅内假体，它的手术风险性和后遗症是可想而知的。目前的研究报道表明视皮层假体往往容易导致颅内感染和潜在的排异反应，诱发局部癫痫。同时，由于缺乏视网膜和丘脑的等部位的前期视觉处理信息，因而视皮层假体能提供的幻视点空间分辨率很低。另外，由于植入的视觉假体不能也不可能覆盖所有视皮层区域，皮层假体为患者提供视野也十分有限。

3.2 视神经假体优缺点

以目前的技术而言，视网膜假体和视皮层假体的电刺激不可能覆盖整个视觉区域。然而，视神经却在相对小的区域内包含了整个视野的信息。因此，从理论上说，通过视神经假体的电极刺激视神经，是可以使患者 在整个视野里获得光感，所以视神经假体的应用前景很值得期待。此外，与视皮层假体相比，视神经假体植入的手术风险较小，引起的并发症较轻，可以使用较低的电流进行刺激，减少电流对周围神经组织的损伤。但是视神经假体的实际应用也还存在着一定的局限性：首先，具有完整功能的视神经是视神经假体植入的前提，因此视神经假体无法应用于视神经受损的患者；其次，神经内刺激微电极会对视神经造成机械损伤；最后，视觉和视神经刺激确切的对应关系还不明确，实施选择性刺激需要较多的刺激电极，从而进一步增加了损害视神经的危险。此外，就目前研究来看，视神经束的尺寸限制了电极的数目，视神经假体空间分辨率仍然有限，因此刺激电极的设计和开发显得尤为重要。

3.3 视网膜假体优缺点

在所有视觉假体中，视网膜植入体出现相对较晚。它可以分为视网膜上假体和视网膜下假体两大类。尽管视网膜修复技术只适用于视网膜变性疾病，但对于最常见的眼部疾病视网膜色素变性和老年黄斑变性，它可以利用视网膜中大量的存活细胞，存在诸多优点。但是，不管是视网膜上假体还是视网膜下假体，现在在实际研究中也都存在着一些不尽人意之处。

对于视网膜下假体而言：首先，它只能适用于光感受器以后所有视觉通路都没有受损的患者；其次，假体植入后可能会阻止了从脉络膜向剩余视网膜组织营养的输送通路，会造成剩余结构的萎缩；最后，手术过程复杂，存在一定的风险。此外，在技术方面，目前的光电二极管阵列分辨率和灵敏度低都还比较低。

而对视网膜上假体而言：首先手术对刺激电极和视网膜的接触点的对接具有很高的技术要求，且由于接触点可能的移动会引起刺激模式的混乱，因此需要电学方法加以纠正；其次，假体的固定存在一定的困难，很可能导致视网膜裂孔的形成和视网膜脱离。最后，由于视神经节细胞分布较为广泛，刺激器微电极阵列微型化存在一定的困难。此外，电极浸于玻璃体中，易于因为周围电介质环境而短路。

4. 结语与展望

人工视觉假体的研究旨在帮助失明患者在一定程度上重获光明。人工视觉假体根据其植入位置的不同，一般可以分为皮层假体、视神经假体和视网膜假体。视网膜假体又可以分为视网膜上假体和视网膜下假体两大类。这些假体都有着各自的应用前景和局限性。在国外，人工视觉假体的研究起步较早，其取得成果涉及了生理、工程、心理物理等诸多领域，为中国未来在该领域的进一步深入研究提供了宝贵的经验。

参考文献：

［1］P. Hossain，I. W. Seetho，A. C. Browning，ect. Artificial means for restoring vision［J］. *Clinical review*，2005，330（1）：30–33.

［2］G. S. Brindley，W. S. Lewin. The sensations produced by electrical stimulation of the visual cortex［J］. *The Journal of Physiology*，1968，196：479–493.

［3］Q. Ren，X. Chai，K. Wu，ect. Development of C-Sight visual prosthesis based on optical nerve stimulation with penetrating electrode array［J］. *Investigative ophthalmology & visual science*，2008，48（5）：146–404.

作者简介

杨元，男，2010年～2013年在法国巴黎高科高等电信攻读博士学位，博士研究课题为“非侵入式脑机接口技术的大型普适化应用”。该课题是巴黎高科电信、法国国家科学研究中心和法国电信集团的合作课题，同时和巴黎11大、凡尔赛大学机器人研究所、INRIA等保持有密切合作。脑机接口技术是通过测量来自人或动物脑部的信号，并将采集到的信号，通过特征提取和分类，转换成机器指令，以实现大脑和电脑的通信，从而最终达到控制外部设备（鼠标、轮椅或假肢）的目的。目前的研究表明可以利用人脑电波信号帮助被试者实现开灯关灯、打字和控制鼠标等任务。该技术的目前已在助老助残、康复工程等领域有所应用，并开始进入新一代娱乐产业（如美国的Nuerosky公司开发的Mindset系统）。杨元博士从硕士开始就一直在神经科学和康复工程领域开展研究。他的研究涵盖：① 人工视觉假体中的神经生理学研究；② 基于脑电波的脑机接口技术和基于其他生理电信号（眼电）人机交互技术相结合实现多模态人机交互研究；③ 脑机接口技术中的模式识别和信号处理等。其研究成果在一系列国际会议和国际期刊上发表，并先后应邀在法国11大、凡尔赛大学机器人研究所，德国中风与痴呆研究中心做报告。

在法期间，杨元博士曾担任上海交通大学法国校友会秘书长，法国茶与健康协会副会长、秘书长，致公党江苏省委海外联谊工作联络员。是国际电气电子工程协会（IEEE）、国际华人人机交互协会（China CHI）、全法科协、上海市神经科学学会会员。2014年1月起，为全欧神经科学学会联合会（Federation of all European neuroscience societies）会员。目前，担任福建省海外交流协会常务理事、副秘书长。

磁场驱动的水中螺旋形微型机器人

徐天添，Gilgueng Hwang，Nicolas Andreff，St é phane R é gnier ISIR
Institut des Syst è mes Intelligents et de Robotique，Pyramide-T55/65，
BC 173-4 Place Jussieu，75005 Paris

Magnetic Actuated Helical Microswimmers

摘　要：可控的微型机器人在未来的医学中会有很大的应用空间。物体在低雷诺数的流体中的运动需要特殊的技巧。我们通过模仿自然界微生物的运动方式，制作了一种由磁场驱动的螺旋形机器人。在本文中，我们比较了不同几何参数对螺旋新微型机器人的运动性能的影响，从而得出了一种最优化设计。之后我们又提过图像反馈，提出一种对这种螺旋形微型机器人实行闭环控制的可能性。

关键词：微型机器人　螺旋形　磁场驱动

1. 背景介绍

可控的微型机器人在未来的医学中会有很大的应用空间，例如靶向治疗、微创手术等 [1]。微型机器人在流体中的运动等价于在低雷诺数（Reynolds number）下的运动。物体在低雷诺数的流体中的运动比较特殊。宏观的“游泳”技巧很难应用到微观的“游泳”中去 [2]。如何在微观世界中“游泳”，我们可以参考自然界中的水生微生物的运动方式。自

然界的微生物在水中有两种运动方式：柔软鞭毛的摆动和螺旋形推进的方式[3]。通过模仿自然界微生物的运动方式，如图1所示，我们可以有以下三种做水中微型机器人的理念。

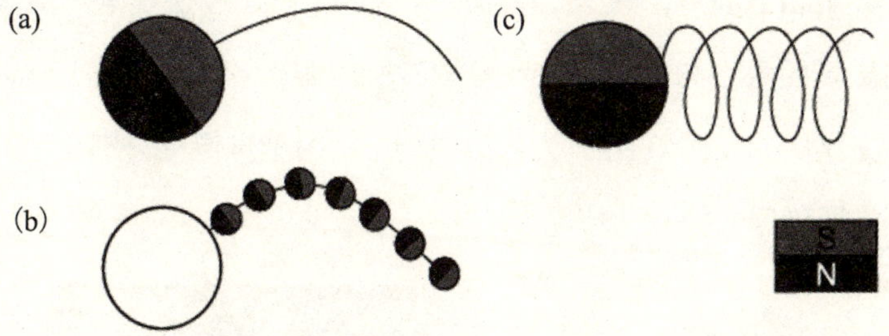

（a）机器人由一条柔软的尾巴加一个有磁性的头组成，放置在一个震动的磁场中，使得柔软的尾巴摆动，从而使机器人前进。

（b）机器人的尾巴由多个有磁性的物体串联而成，同样放置在震动的磁场中，使之前进。

（c）机器人有一个有磁性的头和一个螺旋形的尾巴组成，放置在旋转的磁场中，从而螺旋形前进。

图1　三种水中微型机器人的理念

这三种机器人的理念，第三种可控性最强，可以通过控制磁场的频率和方向来决定机器人的速度和前进方向。我们决定采用这个理念来制作水中的微型可控机器人。

从1996年起，几个实验室都有科学家制作了不同的磁场驱动的水中螺旋形微型机器人，并进行了水中运动的实验[4]~[9]。这些水中螺旋形微型机器人和几何参数（半径、步距、宽度）各不相同。我们会在本文中就各个参数对螺旋形微型机器人运动性能的影响进行分析，得出对运动性能影响最大的参数，从而得到螺旋形微型机器人参数的最优化设计。

2. 系统介绍

就我们研究的磁场驱动的螺旋形机器人而言。螺旋形机器人有一个磁性的头，由一个旋转的均匀磁场控制。这个旋转的均匀磁场由三个相互垂直排列的亥姆霍兹线圈来实现，如图2所示。我们通过控制电流的大小和频率控制磁场的大小和频率。在实验中，所用的磁场的大小为10 mT，磁场的频率变化从0 Hz到10 Hz，为低频磁场。由两个摄像机分别在上面和旁边记录螺旋形机器人的运动情况。

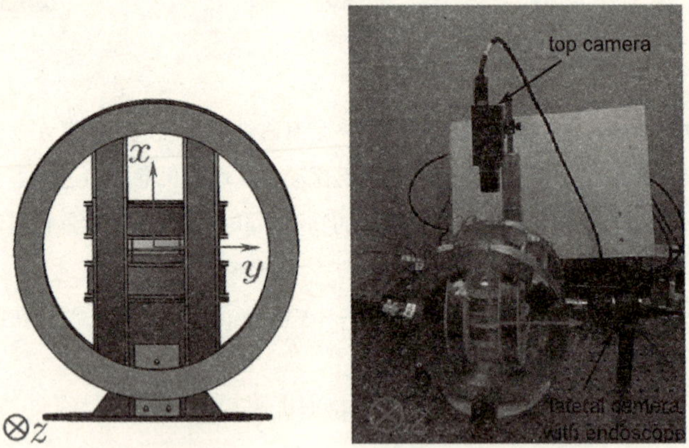

（a）CAD模型图　　（b）实物图

图2　三个相互垂直排列的亥姆霍兹线圈

3. 几何参数对螺旋形机器人运动性能的影响

为了比较各个参数对螺旋形微型机器人运动性能的影响，我们制作了不同几何参数的螺旋形机器人。几何参数遵循3个参数、2个等级的试验设计（2^3 factorial design of experiments），如图3所示。螺旋形机器人由3D快速成型打印技术制作，打印材料为树脂。一个小的圆形磁铁片嵌在螺旋形机器人前面的圆柱形凹槽里。通过用磁场控制磁铁头的旋转来控制螺旋形机器人的前进速度。

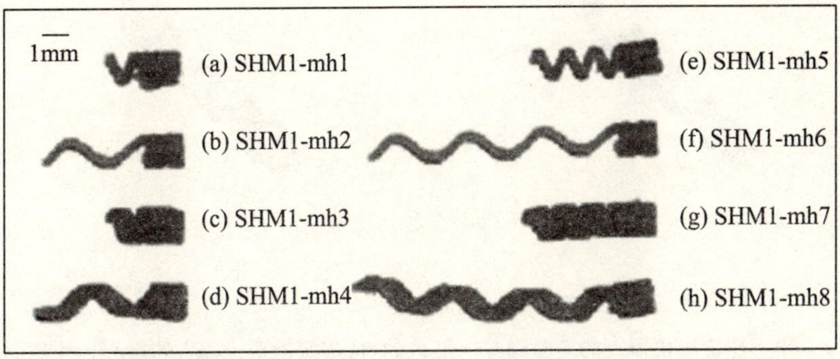

图3　我们制作的不同几何参数的螺旋形机器人。几何参数遵循3个参数、2个等级的试验设计（2^3 factorial design of experiments）

我们比较了这8种不同参数的螺旋形机器人的在相同磁场大小和频率下的前进速度。通过比较，得出对运动性能影响力最大的参数是螺旋形机器人的步距。步距越大的螺旋形机器人，运动性能越好。螺旋形尾巴的宽度会影响机器人的坚固程度。例如SHM1-mh6型机器人虽说运动性能与SHM1-mh8型机器人相当，但是由于尾巴宽度太小，在使用过程中非常容易损坏。于是在之后的研究中，我们采用了SHM1-mh8型号的螺旋形机器人。

4. 图像反馈的闭环控制

我们希望通过图像反馈，实现一个闭环控制。通过ViSP[10]，我们对螺旋形机器人进行实时跟踪。借助于图像矩（Image moment）的概念，我们可以实时计算螺旋形机器人的重心和中心轴，如图4所示。

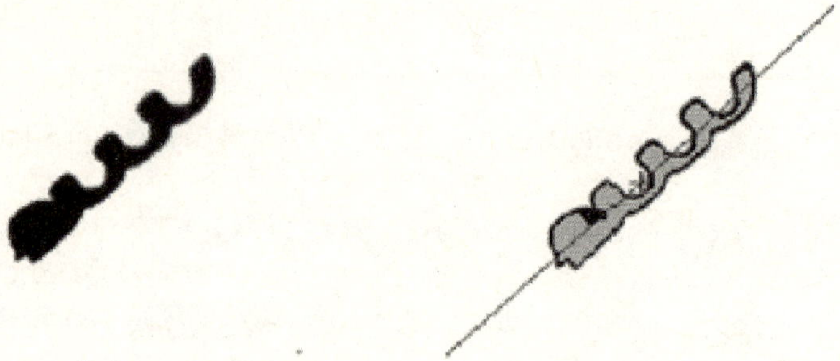

（a）　　　　　　　　　　　　（b）

（a）摄像机实时记录的螺旋形机器人

（b）螺旋形机器人的实时跟踪。红色的+代表螺旋形机器人的重心。红色的直线代表螺旋形机器人的中心轴。

图4　通过ViSP, 实时跟踪螺旋形机器人

对磁场驱动的螺旋形机器人的控制，我们采用路径跟踪的模式（Path following）。我们给定一个参考路径C（s），然后控制两个误差，使其分别趋近于零。其一为距离误差，是螺旋形机器人的重心到参考路径的最小距离。其二为角度误差，是螺旋形机器人的中心轴和参考路径上离螺旋形机器人重心最近的那一点的切线的角度差。对控制法则的建模和分析，就是我们研究的方向。

参考文献

［1］B.J. Nelson，I.K. Kaliakatsos，J.J. Abbott. Microrobots for minimally invasive medicine［J］. *Annual Review of Biomedical Engineering*，2010，12（1）：55–85.

［2］E.M. Purcell. Life at low reynolds number［J］. *American Journal of Physics*，1977，45（1）：3–11.

［3］H.C. Berg，R.A. Anderson［J］. Bacteria Swim by Rotating Their Flagellar Filaments. *Nature*. 1973，245（5425）：380–382.

［4］T. Honda，K.I. Arai，K. Ishiyama. *Micro Swimming Mechanisms*

Propelled by External Magnetic Fields. Magnetics [J] . *IEEE Transactions*, 1996, 32 (5): 5085 –5087.

[5] D.J. Bell, S. Leutenegger, K.M. Hammar, et al. Flagella-like Propulsion for Microrobots Using a Nanocoil and a Rotating Electromagnetic Field [J] . *In Robotics and Automation*, *2007 IEEE International Conference on*, 2007: 1128–1133.

[6] L. Zhang, J.J. Abbott, L. Dong, et al. Artificial Bacterial Flagella: Fabrication and Magnetic Control [J] . *Appl. Phys. Lett.* 2009, 94 (6).

[7] G. Hwang, R. Braive, L. Couraud, et al. Electro-Osmotic Propulsion of Helical Nanobelt Swimmers [J] . The International Journal of Robotics Research, 2011, 0 (7): 806–819.

[8] S. Tottori, L. Zhang, F. Qiu, et al. Magnetic Helical Micromachines: Fabrication, Controlled Swimming, and Cargo Transport [J] . *Advanced Materials*, 2012, 24 (6): 811–816.

[9] T. Xu, G. Hwang, N. Andreff, et al. Modeling and Swimming Property Characterizations of Scaled-up Helical Microswimmers [J] . *Mechatronics*, *IEEE/ASME Transactions on*, 2013, (99): 1–11.

[10] E. Marchand, F. Spindler, F. Chaumette. Visp for Visual Servoing: a Generic Software Platform with a wide Class of Robot Control Skills [J] . *Robotics Automation Magazine*, IEEE, 2005, 12 (4): 40 –52.

作者简介

徐天添　1987年3月16日，北京

留法求学经历：10/2010— ISIR Institut des Systèmes Intelligents et de Robotique，UPMC Universite de Pierre et Marie Curie / 巴黎六大机器人与人工智能实验室

准备学位：工学博士　博士课题：*Conception and Control of Bio-inspired Helical Microswimmers*

2009—2010 Ecole Centrale Paris（Master Degree：Industry Engineering）

巴黎中央理工 硕士学位：工业工程

2007—2010 SUPMECA Paris（Engineering Degree：Mechanics）

巴黎高等机械学院 工程师学位：机械

2005—2007 MPSI/MP* LYCEE FAIDHERBE Lille

里尔FAIDHERBE中学数理预科班

国内母校：1999—2005 北京师范大学附属实验中学

发表论文：

[1] Xu. T.; Hwang, G.; Andreff, N.; Régnier, S., (2012). *Scaled-Up Helical Nanobelt Modeling and Simulation at Low Reynolds Numbers*. ICRA'12 IEEE International Conference on Robotics and Automation, pp. 4045-4051.

[2] Xu. T.; Hwang, G.; Andreff, N.; Régnier, S., "*Modeling and Swimming Property Characterizations of Scaled-Up Helical Microswimmers*," Mechatronics, IEEE/ASME Transactions on , vol.PP, no.99, pp.1, 11, 2013

[3] Xu. T.; Hwang, G.; Andreff, N.; Régnier, S., "*The Rotational Propulsion Characteristics of Scaled-Up Helical Microswimmers with Different Heads and Magnetic Positioning*", IEEE/ASME International Conference on Advanced Intelligent Mechatronics (AIM2013) (Best Paper Finalist Award)

基于风险因素的工业系统绩效测量体系研究

李　璠

法国国立高等工程技术大学校，设计、制造工程与控制实验室

4，rue Augustin Fresnel，57078 Metz Cedex 3，France

摘　要：绩效测量是工业工程和管理科学领域的核心研究课题之一，一直被企业管理人员和相关学科的研究人员所关注。本课题拟研究建立一种基于风险的绩效测量体系，从而帮助工业系统的评估人员能够在不确定的、相互矛盾的条件下模拟系统各个利益相关者的期望目标，随后从一系列能将系统价值与利益相关者价值相统一的备选方案中筛选出一种最佳解决方案。

关键词：绩效测量　决策支持　风险管理

1. 研究背景

从20世纪60年代起，绩效测量理论得到了不断的发展和完善。最初的绩效测量理论仅仅考虑系统成本一项因素[1]。然后该理论扩展为基于对系统财务性绩效指标和非财务性绩效指标的测评[2]。最近，绩效测量理论与绩效管理系统相结合，依靠多项指标对系统绩效的多维性质进行分析研究。然而通过对现有绩效测量与管理工具及理论的评审和分析，发现当前没有任何一种工具或理论能够全面处理已有的绩效概念。除此之外，全球化竞争和商业革命也在不断改变应被研究考虑的指标。

在这一背景下，本课题拟研究建立一种基于风险因素的绩效测量体系，以帮助工业系统的评估人员能够在不确定的、相互矛盾的条件下模拟各个利益相关者的期望目标，进而从一系列能将系统价值与利益相关者价值相统一的备选方案中筛选出一种最佳解决方案。

2. 基本模型

绩效测量体系通常被当作一种管理工具应用于评估、分析、指导和改善企业作业流程。绩效指标作为绩效测量系统的基本元素，被用于表达一个企业的战略以及评估企业实际作业流程与其目标的一致性。在本研究中，风险因素的定义是对于实现系统目标而要面临的各种风险的估计，也被看作是各种基本风险的集合。

2.1 目标模型

系统各个利益相关者的价值是其期望目标，可通过价值中心法对其进行分析[3]。使用该方法可确定所研究作业流程的各个目标O_i（$i = 1$，… n）并定义与之相应的测量方法M_j（$j = 1$，…，m）。例如在供应链流程中，最高目标是“满足客户订单要求”，次一级目标则是对上一级目标的具体表达，例如满足客户订单要求可以表述为：“按时配送”“最低成本”和“高质量”。以此方法对系统每一级目标进行分解表达，如图1所示。

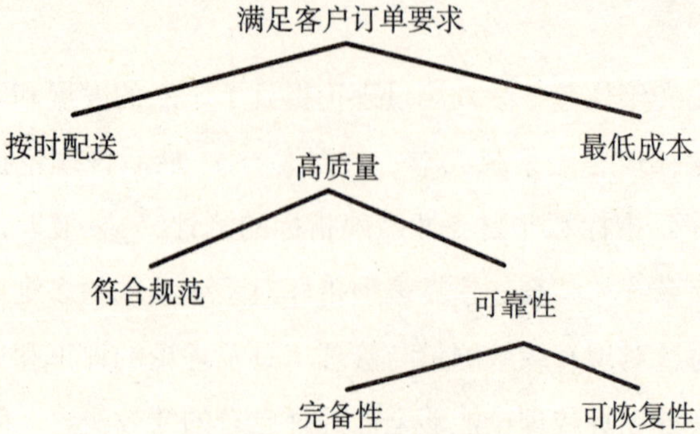

图1　系统目标分解图

2.2 活动模型

活动模型是基于风险因素的绩效测量体系的一部分，并且直接或间接支持系统某些目标的实现[4]。为构建该模型，首先要定义系统中的各项活动，其次再按照作业流程确定各项活动的先后顺序。在生产制造流程中，按照从需求分析到产品制造特性的顺序定义各项活动[5]；在供应链流程中，从一般作业流程模式开始定义各项活动[6]。在上述定义完成之后，使用IDEF3图表示各项活动的先后顺序、逻辑关系和约束条件，如图2所示。

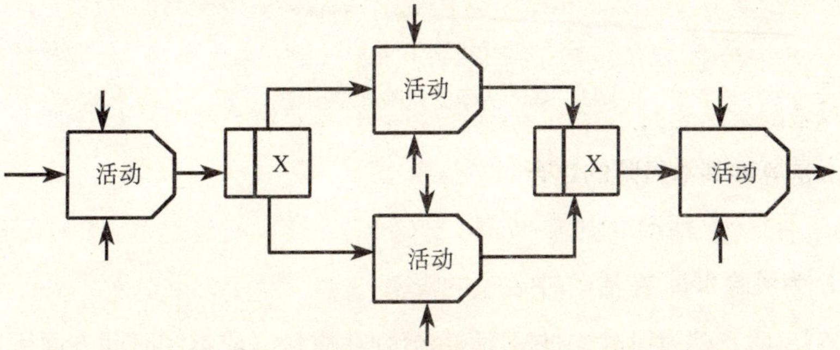

图2　基于IDEF3的活动模型示意图

2.3 风险模型

与目标模型类似，通过价值中心法对系统的风险因素进行分解，确定各级风险。然后结合在活动模型中定义的各项活动，定义风险/活动矩阵，如表1所示。

表1　风险/活动矩阵

风险（R）/活动（A）	R1	R2	R3	R4	R5
A1		X		X	X
A2	X		X		

2.4 基于活动的风险评估模型

在该模型中，作业流程中的每项活动与一个或多个风险因子RF_i（$i=$

1，$\cdots n$）相关联，风险因子可以引发相关的风险事件，从而通过有关活动影响系统目标的实现，如图3所示。

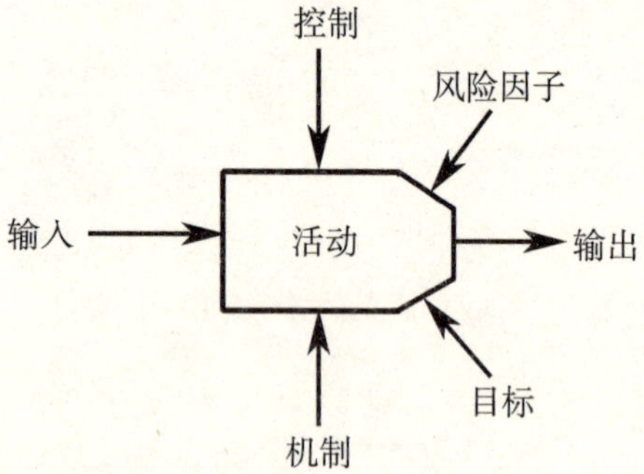

图3　基于活动的风险评估模型

活动i的风险计算公式为[7]

$$\text{活动}i\text{的风险}=P_{ij}\left(C_{ij}^{q}+C_{ij}^{c}+C_{ij}^{t}\cdots\right) \quad (1)$$

P_{ij}为风险事件j在活动i中发生的概率；

C_{ij}^{q}，C_{ij}^{c}，C_{ij}^{t}为风险事件j对活动i分别在质量、成本和时间方面所造成的影响。

活动i的整体风险R_i为

$$R_i=\sum_{j=1}^{J}\left(P_{ij}\times C_{ij}^{obj}\right) \quad (2)$$

但是每项风险事件在同一系统中的重要性是不同的。某些风险对系统目标造成的影响比其他风险更为显著。所以在此基础上为每项风险增加权重系数d_{ij}，则活动i的整体风险R_i的计算公式改进为

$$R_i=\sum_{j=1}^{J}d_{ij}\left(P_{ij}\times C_{ij}^{obj}\right) \quad (3)$$

路径P_k的风险为

$$R\left(p_k\right)=\sum_{\forall i\in pk}R_i=\sum_{\forall i\in pk}\sum_{j=1}^{J}d_{ij}\left(P_{ij}\times C_{ij}^{obj}\right) \quad (4)$$

其中，$\sum_{k=1}^{K}P_r\left(P_k\right)=1$

综上所述，由路径K所组成的作业流程P的总体估算风险为

$$E\left(R_p\right)=\sum_{k=1}^{K}P_r\left(P_K\right)\left(\sum_{\forall i\in pk}\sum_{j=1}^{J}d_{ij}\left(P_{ij}\times C_{ij}^{obj}\right)\right) \quad (5)$$

3. 小结与展望

本研究介绍了一种基于风险因素的工业系统绩效测量体系。该体系中包含了风险因素的定义和模型以及数学计算公式，为基于风险因素的工业系统绩效测量和决策分析提供了相关理论依据。

该课题未来的研究方向主要包括：在现有理论基础上，增加价值、成本和收益三项因素，依靠多项指标对系统绩效的多维性质进行分析研究，从而完善工业系统绩效测量和管理框架体系，更好地辅助工业系统决策制定。

参考文献：

[1] Johnson，H.T.，Kaplan，R.S. Relevance lost：The rise and fall of management accounting [M]. Harvard Business School Press. 1987.

[2] Dossi，A.，Patelli，L. You learn from what you measure：Financial and non-financial performance measures in multinational companies [J]. Long Range Planning. 2010，43：498–526.

[3] Keeny，R.L. Value-Focused Thinking：A Path to Creative Decision-making. Cambridge：Harvard University Press，1996.

[4] Curtis，B.，Kellner，M.L.，Over，J. Process modeling [J]. Commun. ACM. 1992，35：75–90.

[5] Sormaz，D.N.，Khoshnevis，B. Generation of alternative process plans in integrated manufacturing systems [J]. Journal of Intelligent Manufacturing. 2003，14：509–526.

[6] SCC，SCOR v11.0. Supply Chain Council，Inc，Washington. 2013.

[7] Larson，N.，Kusiak，A. Managing design processes：A risk assessment approach [J]. IEEE Transactions on Systems，Man，and Cybernetics–Part A：Systems and Humans. 1996，26：749–759.

作者简介

李璠，男，汉族，中共党员

出生日期：1986年5月21日

籍贯：山西省祁县

教育背景：

2004.09—2008.07　同济大学，交通运输专业，本科，期间担任交通运输工程学院团委助理

2008.09—2011.02　同济大学/法国国立桥路大学校，交通运输规划与管理专业/供应链设计与优化专业，硕士研究生（中法联合培养，双硕士学位）

2013.09—今　法国国立高等工程技术大学校，工业工程专业，博士研究生

研究方向：

供应链设计与优化、运输物流系统建模与仿真、决策支持、绩效测量、风险管理

论文发表：

1. 李璠，“DRT模式公交巴士服务探讨”，科学之友，2011年第4期，P120-121，2011年2月
2. 李璠，“基于出行需求的公交巴士服务构建初探”，江西科学，第29卷，P244—247，2011年3月
3. 李璠，“关于公交巴士服务的新模式”，山西道路运输，2011年第2期，P38—40，2011年3月

课题研究：

1. 应对2010年上海世博会铁路客运组织研究（2006.10—2007.04）

该项目研究成果荣获同济大学第六届同路人交通科技大赛三等奖，2006年上海高校学生创造发明“科技创业杯”三等奖。

获奖情况：

2009.06　同济大学法语演讲比赛二等奖

2007.12　2006—2007学年同济大学优秀学生干部

2007.12　2006—2007学年同济大学社会活动奖学金

2007.01　2006年同济大学大学生暑期社会实践先进个人

2006.12　2005—2006学年同济大学社会活动奖学金

2006.01　2005年度上海市政府奖学金

2005.12　2004—2005学年同济大学社会活动奖学金

Pattern Formation Modeling for Thin Films on Soft Substrates

Fan Xu（徐凡）

Laboratoire d' Etude des Microstructures et de Mécanique des Matériaux, LEM3, UMR CNRS 7239, Université de Lorraine, Ile du Saulcy, 57045 Metz Cedex 01, France.

Abstract: Wrinkles of a stiff thin layer attached on a soft substrate have been widely observed in nature (see Fig. 1) and these phenomena have raised considerable interests for several decades. This paper aims at introducing brief ideas of our research on applying advanced numerical methods for pattern formation modeling of film/substrate systems.

Keywords: Wrinkling; Post-buckling; Bifurcation; Perturbation Technique; Finite Element Method.

Thin films attached on soft substrates are important for electronic systems that require or benefit from mechanical stretchability, such as flexible displays, electronic eye camera, conformable skin sensors, smart surgical gloves, and structural health monitoring devices [1]. When a hard film is deposited on such a soft material, subject to a residual compressive stress due to the large thermal expansion mismatch coefficients between the film and substrate during cooling process, the film may form a pattern of wrinkles and remain bonded to the substrate [2]. The wrinkles are a nuisance in some applications, but may be used as stretchable interconnects, as templates for device fabrication, or as a means to evaluate mechanical properties of materials.

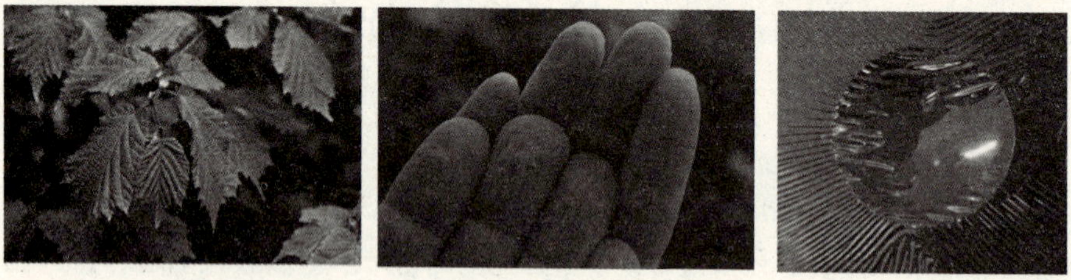

Figure 1 The left shows leaf wrinkles and the middle demonstrates finger wrinkles (photos from internet); the right is wrinkled circle in thin Kapton membrane (photo from University of Cambridge)

In terms of pattern formation study, several theoretical and experimental works have been devoted from linear perturbation analysis to nonlinear stability analysis. However, most previous studies have been mainly constrained to determine the critical conditions of instability and the corresponding wrinkling patterns near the instability threshold. The post-buckling evolution and mode transition of surface wrinkles are only recently being pursued [3, 4]. Can one describe the whole evolution of buckling and post-buckling of this system? Under what loading can each type of pattern be observed? What are their amplitudes and wavelengths? These questions will be addressed in our research.

Our research work aims at applying advanced numerical methods for pattern formation modeling of film/substrate system from 2D to 3D, from classic model to multi-scale model, and focuses on the post-bifurcation evolution involving secondary bifurcations and advanced modes. For this purpose, finite element (FE) models based on the Asymptotic Numerical Method (ANM) [5, 6] are developed for nonlinear analysis of wrinkle formation [7-11]. Dimensional analysis is performed to find characteristic parameters that influence instability patterns. In the 2D model, the film undergoing moderate deflections is described by Föppl-von Kármán nonlinear elastic theory, while the substrate is considered to be a linear elastic solid. Following the same strategy, we extend the 2D work to 3D situation by coupling shell elements representing the film and block elements describing the substrate. Therefore, the spatial distribution of wrinkling modes like stripe or checkerboard can be investigated.

To solve the resulting nonlinear equations, we adopted the ANM which appears as a significantly efficient continuation technique without any corrector iteration. The underlying principle of the ANM is to build up the nonlinear solution branch in the form of relatively high order truncated power series. Since few global stiffness matrix inversions are required (only one per step), the performance in terms of computing time is quite attractive. Moreover, unlike incremental-iterative methods, the arc-length step size in the ANM is fully adaptive since it is determined a posteriori by the algorithm. A small radius of convergence and step accumulation appear around the bifurcation and imply its presence. Furthermore, a bifurcation indicator well adapted to the ANM, is computed to detect the exact bifurcation points. This indicator is a determinant of a reduced stiffness matrix, which measures the intensity of the system to perturbation forces. By evaluating this indicator through an equilibrium branch, all the critical points existing on this branch and the associated wrinkling modes can be determined.

Numerical results reveal different types of wrinkling modes under different loading and boundary conditions (see Fig. 2). The evolution of patterns and advanced post-bifurcation mode including period-doubling and localized mode have been observed in the post-buckling range. The results are expected to provide insight into the formation and evolution of wrinkle patterns in film/substrate systems and be helpful to control the surface morphology, which could apply in the pattern formation for micro-and nano-fabrication, biocompatible topographic matrices for cell alignment, force spectroscopy in cells, tunable phase optics and high precision micro-and nano-metrology methods.

Figure 2 Numerical calculation results on different types of wrinkling patterns.

References

[1] S. Wang, J. Song, D.H. Kim, et al. Local versus global buckling

of thin films on elastomeric substrates. *Applied Physics Letters*, 93: 023126, 2008.

[2] N. Bowden, S. Brittain, A.G. Evans, J.W. Hutchinson, G.M. Whitesides. Spontaneous formation of ordered structures in thin films of metals supported on an elastomeric polymer. *Nature*, 393: 146–149, 1998.

[3] F. Brau, H. Vandeparre, A. Sabbah, C. Poulard, A. Boudaoud, P. Damman. Multiple-length-scale elastic instability mimics parametric resonance of nonlinear oscillators. *Nature Physics*, 7: 56–60, 2011.

[4] J. Zang, X. Zhao, Y. Cao, J.W. Hutchinson. Localized ridge wrinkling of stiff films on compliant substrates. *Journal of the Mechanics and Physics of Solids*, 60: 1265–1279, 2012.

[5] B. Cochelin, N. Damil, M. Potier-Ferry. Asymptotic-numerical methods and Padé approximants for non-linear elastic structures. *International Journal for Numerical Methods in Engineering*, 37: 1187–1213, 1994.

[6] B. Cochelin, N. Damil, M. Potier-Ferry. Méthode asymptotique numérique. Hermès Science Publications, 2007.

[7] F. Xu, S. Belouettar, M. Potier-Ferry, H. Hu. A finite element model for wrinkling analysis of thin films on compliant substrates. 12*th U.S. National Congress on Computational Mechanics* (*USNCCM*12), Raleigh, North Carolina, U.S.A., July 22–25, 2013.

[8] F. Xu, S. Belouettar, M. Potier-Ferry, H. Hu. Surface wrinkling of a stiff thin film on a functionally graded substrate. *International Conference on Science and Technology of Heterogeneous Materials and Structures* (ICSTHMS 2013), Wuhan, P.R. China, October 11–13, 2013.

[9] F. Xu, S. Belouettar, M. Potier-Ferry. Pattern formation finite element modeling for thin films on soft substrates. 11*th World Congress on*

Computational Mechanics（*WCCM XI*）& *5th European Conference on Computational Mechanics*（*ECCM V*），Barcelona，Spain，July 20–25，2014.

［10］F. Xu，S. Belouettar，H. Hu，M. Potier-Ferry. A novel nonlocal reduction-based coupling approach via Arlequin method on instability patterns. *Proceedings of 11ème Colloque National en Calcul des Structures*（*CSMA* 2013），Giens，Var，France，May 13–17，2013，8 pages.

［11］F. Xu，H. Hu，S. Belouettar，M. Potier-Ferry. Analysis of reduced coupling and prolonged coupling approach via Arlequin method on instability pattern. *Proceedings of the 6th International Conference on Coupled Instabilities in Metal Structures*（*CIMS* 2012），Glasgow，UK，December 3–5，2012，ed. J. Loughlan，87–94.

A Short Survey on Hamiltonian Line Graphs

Hao Li, Weihua Yang

Hao Li is a Directeur de Recherche in CNRS (France) and a Changjiang Lectureship Chair Professor (China).

Laboratoire de Recherche en Informatique

UMR 8623, CNRS.-Université de Paris-sud

91405 Orsay, France.

Institute for Interdisciplinary Research

Jianghan University

Wuhan 430019, China

Weihua Yang was a PhD student under the guidance of Hao Li from Dec. 10th, 2010 to Sep. 27th, 2013 at LRI-CNRS Paris-Sud University and is a professor in Taiyuan University of Technology.

Department of Mathematics

Taiyuan University of Technology

Shanxi Taiyuan 030024, China

Abstract: In this short survey, we mainly introduce our works on the hamiltonicity of line graphs.

1. Introduction

A *hamiltonian path* is a path that visits each vertex exactly once. A graph that contains a hamiltonian path is called a traceable graph. A graph is hamiltonian-connected if for every pair of vertices and between these two vertices there is a hamiltonian path. A graph that contains a hamiltonian cycle is called a hamiltonian graph.

In 1856, Hamilton invented a mathematical game, that is the Icosian Game. In a dodecahedron, each of its twenty vertices was labeled with the name of a city and the problems is to find a hamiltonian cycle in this dodecahedron graph, to make a voyage around the world. Hamiltonian paths and cycles are named after William Rowan Hamilton who invented the Icosian Game. The hamiltonian problem, determining when a graph contains a hamiltonian cycle, has long been fundamental in graph theory. Determining whether hamiltonian cycles exist in graphs is NP-complete. Therefore it is natural and very interesting to study sufficient conditions for hamiltonicity.

On the hamiltonian problem, one may find many well-known theorems in any graph theory textbook. So it is not necessary and impossible to give a detailed survey here. In particular, Li recently in [12] surveyed the results due to Dirac's theorem which started a new approach to develop sufficient conditions on degrees for a graph to be Hamiltonian. For a survey on the hamiltonian problem on Cayley graphs, we refer to [20] by Witte and Gallian. Gould gave two nice surveys in [7, 8] which contains many problems on generalizations of hamiltonian problem; Bauer *et al.*[3] gave a survey which focus on the toughness of graphs and the hamiltonian problem. We suggest the readers to refer to the surveys for different topics of the hamiltonian problem[4, 5, 15]. In this short survey, we shall introduce our works on hamiltonicity of line graphs.

One of the important topics in hamiltonian graph theory is the hamiltonicity of claw-free graphs (*i.e.*, a graph is called claw-free if it contains no induced claw, K1, 3). Before we introduce our main results, we start by presenting the following conjecture due to Matthews and Sumner.

Conjecture 1 (Matthews and Sumner[14]). *Every 4-connected claw-free graph is hamiltonian.*

It is well known that every line graph is claw-free. Thomassen in 1986 posed another conjecture as follows:

Conjecture 2 (Thomassen[18]). *Every 4-connected line graph is hamiltonian.*

Clearly, Conjecture 1 is stronger that Conjecture 2. Herbert Fleischner asked whether they are equivalent? To answer the question, Ryjacek introduced a closure operation (It is based on adding edges without destroying the (non) hamiltonicity (similar to the Bondy-Chvatal's closure for graphs with nonadjacent pairs with high degree sums). The edges are added by looking at a vertex v and the subgraph of G induced by N(v) (neighborhood of v in G). If G[N(v)] is connected and not a complete subgraph, all edges are added into G[N(v)] (subgraph induced by N(v)) to make it a clique (complete subgraph). This procedure is repeated in the new graph, etc., until it is impossible to add any more edges. The resulting graph is called R-closure of G, denoted by cl(G). By using the R-closure, Ryjacek[17] proved the following

Theorem 1 ([17]). *Let G be a claw-free graph. Then*

(1) *the closure cl(G) is uniquely determined*;

(2) *cl(G) is hamiltonian if and only if G is hamiltonian*;

(3) *cl(G) is the line graph of a triangle-free graph.*

For Conjecture 2, the first nice result was due to Zhan [23], as follows.

Theorem 2 ([23]). Every 7-connected line graph is hamiltonian-connected.

By using R-closure, Ryjacek generalized the theorem above.

Theorem 3 ([17]). Every 7-connected claw-free graph is hamiltonian.

The first author showed that every 6-connected claw-free graph with at most 34 vertices of degree 6 is Hamiltonian (*ref*: A note on hamiltonian claw-free graphs, Rapport de Recherche, LRI, No. 1022, Univ. de Paris-Sud, 1996). Very recently, an important progress was obtained by Kaiser and Vrana [9].

Theorem 4 ([9]). Every 5-connected line graph (claw-free) with minimum degree at least 6 is hamiltonian.

2. **Hamiltonicity of line graphs**

So far, several different topics are mentioned on hamiltonian problem of line graph (*i.e.*, claw-free graph). The following one lead to our research mentioned below. Lai *et al.* in [10] considered the following problem.

Question 1. *For 3-connected line graphs, can high essential connectivity guarantee the existence of a hamiltonian cycle?*

Question 2. *Is every 3-connected essentially 4-connected line graph is hamiltonian?*

For the first question, they proved the following theorem.

Theorem 5 (Lai *et al.* [10]). *Every 3-connected, essentially 11-connected line graph is hamiltonian.*

We strengthened the theorem above to hamiltonian connected by using the reduction method introduced by Catlin in [6].

Theorem 6 ([21]). *Every 3-connected essentially 11-connected line graph* (*claw-free graph*) *is hamiltonian connected.*

Later, we consider the second problem by using a theorem of Nash-Williams[16] and Tutte[19].

Theorem 7 ([13]). *Every* 3-*connected essentially* 10-*connected line graph* (*claw-free graph*) *is hamiltonian connected.*

For the first question, we obtained a sharp conditions below.

Theorem 8 ([22]). *Let L* (*G*) *be a* 3-*connected, essentially* 4-*connected line graph. If G has at most* 10 *vertices of degree* 3, *then L* (*G*) *is either hamiltonian, or G is contractible to the Petersen graph.*

The results above are based on a well-known result that if a graph has a spanning trail, then its line graph is hamiltonian. Suppose a graph G contains k-edge disjoint spanning closed trails. One may ask the following problem: How many edge-disjoint hamiltonian cycles are in L (G) ? We partially answer the problem.

Theorem 9 ([1]). *If the minimum degree of G is at least 4k and G has k edge-disjoint spanning closed trails, then there are k edge-disjoint Hamilton cycles in L* (*G*).

Next, we introduce one theorem on the hamiltonian cycles in spanning subgraph of line graphs. Ryjacek's closure turns a claw-free graph G to a line graph by adding edges without changing the hamiltonicity of G. Thus, the hamiltonian line graphs have more edges than the hamiltoncicity required. So we try to look for some rules to remove edges from a hamiltonian line graph as many as possible, and without changing the hamiltonicity of the resulting graph. We define semi-line graph of G with vertex set E (G), and every vertex e= υ ν (υ and ν are two end vertices of the edge e in G) is adjacent to at least min{d (u) −1, ⌈3d (u) /4+1/2⌉} vertices of E (u) and also adjacent to at least min{d（v）−1，⌈3d（v）/4+1/2⌉} vertices of E（ν），（where E（u）（E（v），resp.）denotes the incident edges of u （v，resp.）. We showed the following.

Theorem 10 （［2］）. If L（G）is hamiltonian，then the semi-line graphs of G are hamiltonian.

In the end，we list two open problems for the readers.

Question 3. *What is the minimum integer k such that a 3-connected essentially k-connected line graph is hamiltonian（hamilton-connected）?*

Question 4. How many edge-disjoint hamiltonian cycles are in line graph of a graph having k edge-disjoint spanning （dominating）closed trail?

References

［1］Y. Bai，W. He，H. Li，W. Yang，Disjoint hamiltonian cycles in line graphs，submitted for publication.

［2］Y. Bai，W. He，H. Li，W. Yang，On the hamiltonian cycel in the induced sub-graphs of line graph，submitted for publication.

［3］D. Bauer，E.F. Schmeichel，H.J. Veldman. Some recent results on long cycles in tough graphs，Proc. 6th Int. on the theory and application of graphs，Kalamazoo，（1988）113–123.

［4］J.C. Bermond and C. Thomassen，Cycles in digraphs-a survey，J. Graph Theory 5（1981）1–43.

［5］J.A. Bondy，Hamilton cycles in graphs and digraphs.（Proceedings 9th S.E. Conf. on Combin.，Graph Theory and Computing），Congr. Numer. XXI（1978）3–28.

［6］P.A. Catlin，A reduction method to nd spanning Eulerian subgraphs，J. Graph Theory 12（1988）29–45.

[7] R. J. Gould, Updating the hamiltonian problem-A survey, Journal of Graph Theory 15 (1991) 121–157.

[8] R. J. Gould, Advances on the Hamiltonian Problem-A Survey, Graphs and Combinatorics 19 (2003) 7–52.

[9] T. Kaiser, P. Vrana, Hamilton cycles in 5-connected line graphs, European Journal of Combinatorics 33 (2012) 924–947.

[10] H.-J. Lai, Y. Shao, H. Wu, J. Zhou, Every 3-connected, essentially 11-connected line graph is Hamiltonian, Journal of Combinatorial Theory, Series B 96 (2006) 571–576.

[11] H–J. Lai, L. Xiong, H. Yan, J. Yan, Every 3-connected claw-free Z8-free graph is Hamiltonian, Journal of Graph Theory 64 (2010) 1–11.

[12] H. Li, Generalizations of Diracs theorem in Hamiltonian graph theoryA survey, Dis-crete Mathematics, 313 (2.13) 2034–2053.

[13] Hao Li, Weihua Yang, Every 3-connected essentially 10-connected line graph is hamiltonian connected, Discrete Mathematics 312 (2012) 3670–3674.

[14] M.M. Matthews, D.P. Sumner, Hamiltonian results in K1; 3-free graphs, J. Graph Theory 8 (1984) 139–146.

[15] J. Mitchem, E. Schmeichel, Pancyclic and bypancyclic graphs—a survey. Graphs and Applications, (Boulder, Colo., 1982), Wiley-Intersci. Pub., Wiley. New York, (1985) 271–287.

[16] C.St.J.A. Nash-Williams, Edge disjoint spanning trees of nite graphs, J. London Math. Sot. 36 (1961) 445–450.

[17] Z. Ryjacek, On a closure concept in claw-free graphs, J. Combin. Theory Ser. B 70 (1997) 217–224.

[18] C. Thomassen, Re ections on graph theory, J. Graph Theory 10

（1986）309–324.

［19］W.T. Tutte，On the problems of decomposing a graph into n-connected factors，J. London Math. Sot. 36 （1961）231–245.

［20］D. Witte，J. A. Gallian，A survey-Hamiltonian cycles in Cayley graphs. Discrete Math. 51 （1984），293–304.

［21］W. Yang，H. Lai，H. Li，X. Guo，Collapsible Graphs and Hamiltonian Connec-tivity of Line Graphs，Discrete Applied Mathematics 160 （2012）1837–1844.

［22］W. Yang，L. Xiong，H. Lai，X. Guo，Hamiltonicity of 3-connected Line Graphs，Applied Mathematics Letters 25 （2012） 1835–1838.

［23］S. Zhan，On Hamiltonian line graphs and connectivity，Discrete Math. 89 （1991）89–95.

作者简介

Fan Xu（徐凡）

4 Boulevard Arago，57070 Metz，France | fan.xu@univ-lorraine.fr

EDUCATION

2011.09— Present Laboratoire LEM3（UMR CNRS 7239），UNIVERSITE DE LORRAINE METZ

（法国国家科学研究中心7239）

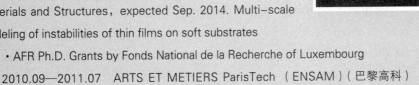

• Doctor of Philosophy（Ph.D.）in Mechanics of Materials and Structures，expected Sep. 2014. Multi-scale modeling of instabilities of thin films on soft substrates

• AFR Ph.D. Grants by Fonds National de la Recherche of Luxembourg

2010.09—2011.07 ARTS ET METIERS ParisTech（ENSAM）（巴黎高科） PARIS

Master of Science（M.S.）in Mechanics of Materials and Structures，M2-Second year top-level master' s degree program in Materials & Engineering Sciences in Paris：MAGIS-Paris

• Degree co-delivered by Arts et M é tiers ParisTech，UPMC（University Paris 6），ENS de Cachan，Ecole Polytechnique，Ecole Centrale de Paris（ECP）

2009.10—2010.06 CONSERVATOIRE NATIONAL DES ARTS ET METIERS（CNAM） PARIS

（法国国立科学技术与管理大学）

Master of Science （M.S.）in Management，cum laude，GPA：3.7/4.0，Top 5%. M2-Second year with specializations in Project Management & Business Engineering

· Ranked 3rd out of 94 for the whole year.

2005.09—2009.06　WUHAN UNIVERSITY（武汉大学）　CHINA

Bachelor of Engineering （B.E.）in Engineering Mechanics，cum laude，GPA：3.7+/4.0，Top 5%

Ranked 2nd out of 40 for four consecutive years.

EXPERIENCE

2011.08—2014.07　Centre de Recherche Public Henri Tudor（卢森堡公共研究中心）　LUXEMBOURG

Researcher as a Ph.D. student involved in the national core project WRINKLE

· Multi-scale modeling and simulation of wrinkling phenomena in nanomaterial and composite structures.

2011.02—2011.07　Laboratoire PIMM （UMR CNRS 8006），ARTS ET METIERS ParisTech　PARIS

（法国国家科学研究中心8006）

Research Internship：Coalescence modeling and experimental validation of sintering of thermoplastic polyamide fibers

· One journal paper （1st author）published in Polymer Materials Science and Engineering（EI）.

2009.01—2009.06　Experimental Research Center of Mechanics，WUHAN UNIVERSITY　CHINA

Graduation Thesis：One-dimensional and two-dimensional multi-scale modeling studies based on the Arlequin method

DISTINCTIONS

2009.12　Winner of Exceptional Bachelor's Degree Thesis of Hubei Province；Top 1%.

2009.06　Winner of honor “Excellent Graduate of Wuhan University”；Top 5%.

2009.06　Winner of First-class paper in the 2nd Science and Technology

Innovation Forum for Civil Engineering Students in Hubei Province; Top 5%.

2005—2008 Winner of Outstanding Student Award, Second-class Scholarship from Wuhan University for three consecutive years; Top 10%.

CONFERENCES

2012—2014 USNCCM12, WCCM XI & ECCM V, CSMA 2013, ICSTHMS 2013, CIMS 2012.

Study on Carbohydrate-Carbohydrate Interaction Involved in Cell Recognition and Adhesion Process

Yongmin Zhang（张勇民）

Sorbonne Université, Institut Parisien de Chimie Moléculaire, CNRS UMR 8232, 4 place Jussieu, 75005 Paris, France

1. **Introduction**

Protein-protein interactions constitute an important mechanism for cell adhesion that controls cellular behaviour. Because most of the cell surface proteins are highly glycosylated, carbohydrate-protein interactions are also involved in cell adhesion and recognition. Recent advances in glycobiology revealed the existence of biologically significant carbohydrate-carbohydrate interactions, and this type of interaction could have a general, fundamental character for cell biology. A typical example is the report from Dr. Hakomori's research group (University of Washington, Seattle) who proposed that carbohydrate-carbohydrate interaction is responsible for the initial step of cell adhesion. Embryogenesis, metastasis, and other proliferation processes are, according to the authors, mediated by carbohydrate-carbohydrate interactions. One of the structures involved in this novel mechanism is the $Lewis^x$ (Le^x) trisaccharide determinant (Gal 1→4 [Fuc 1→3] GlcNAc 1). Le^x is the terminal trisaccharide moiety of numerous cell surface glycolipids (Fig. 1) and glycoproteins involved in selectin-mediated cell-cell adhesion and recognition processes. The interaction between Le^x and Le^x was found to be homotypic, and mediated by the presence of divalent cations such as Ca^{2+}.

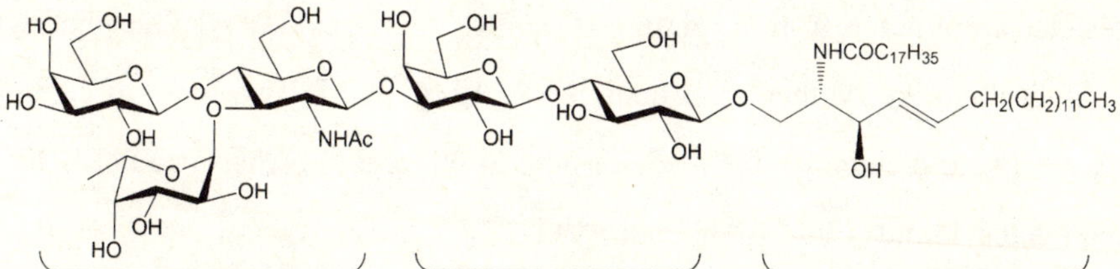

Figure 1 Le^x glycolipid

Since the pioneer report of Hakomori's group on Lex-Lex interaction, several groups have tried to confirm this kind of interactions by using a variety of techniques including nuclear magnetic resonance (NMR) spectroscopy, mass spectrometry (MS), atomic force microscopy (AFM), and surface plasmon resonance (SPR) spectroscopy. Rat basophilic leukaemia cells pre-incubated with purified Le^x-containing glycosphingolipids have been used as a model. Another model system termed "Glycosylated Foldamer" was demonstrated for study of carbohydrate-carbohydrate interaction in terms of individual carbohydrate motifs.

In collaboration with Dr. Perez's group (Ecole Normale Supérieure, Paris), we were able to confirm the existence of this specific interaction by first direct quantitative measurements using a vesicle micromanipulation approach with chemically synthesized natural and non-natural Le^x pentasaccharide glycosphingolipid. The homotypical property of the interaction was also demonstrated by using an isomer of Le^x.

2. **Results and discussion**

2.1 *Direct measurements of Le^x-Le^x interaction using a neoglycolipid.*

Giant vesicles were formed by lipid hydration （after evaporation from chloroform solution）in 320 mOsm sucrose solution. The Le^x functionalized vesicles were made from a mixture of SOPC （StearoylOleoyl Phosphatidyl Choline）and the synthetic Le^x lipid （90：10 by mol）. Two types of control vesicles were prepared：one made of pure SOPC and the other made of SOPC and Lactosyl lipid （90：10 by mol）（Fig. 2）.

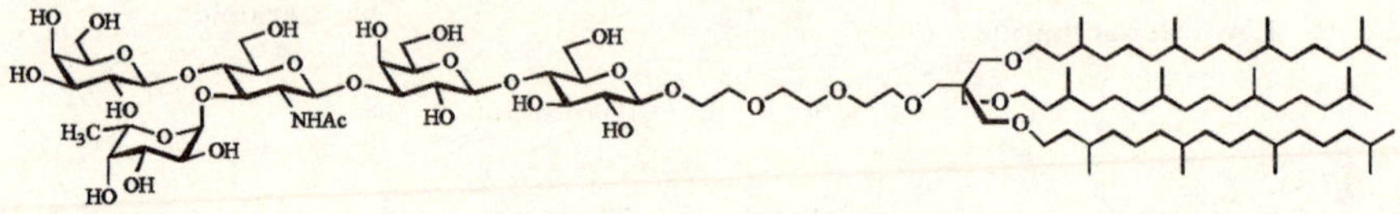

Pentasaccharidic Le^x neoglycolipid

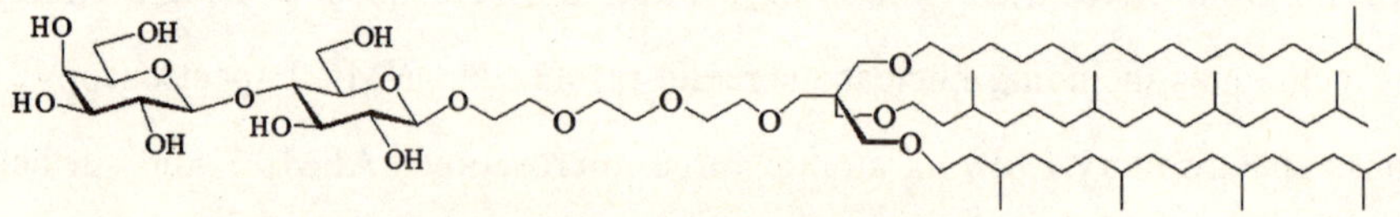

Lactosyl neoglycolipid

Figure 2

The vesicle suspension was added to an aqueous glucose solution chamber of a slightly higher osmolarity （360 mOsm）than that of the vesicles in order to deflate them and make them micromanipulable. Two vesicles were then transferred into another chamber filled with salt solution （either NaCl or $CaCl_2$ at 360mOsm）. Both vesicles were micromanipulated into tangential contact. The contact angle was measured for several tension values of the flaccid vesicle membrane by decreasing the aspiration and then increasing it in order to check the reversibility of the adhesion.

The experiments consisted in comparing the adhesion of two Le^x vesicles in NaCl and in $CaCl_2$ (Le^x / Le^x experiments). As calcium is known to produce sometimes peculiar effects on bilayer interactions, many controls were required. First, it was necessary to compare Le^x / Le^x experiments with experiments in which the Le^x groups from one of the vesicles were absent (Le^x /Lac experiments). Second, it was useful to replace the Le^x groups by another sugar in both vesicles (Lac/Lac experiments). Third, as SOPC is the main component of our vesicles, it was interesting to compare the adhesions obtained with that of pure SOPC vesicles.

Both osmotically controlled vesicles were observed in interference contrast microscopy. One of them was pressurized into a rigid sphere by applying a large bilayer tension, whereas the adherent vesicle was held with low pressure and remained deformable. The adhesion energy Wadh was obtained by determining the contact angle θc of the two vesicles (Fig. 3) and the tension τ_m of the flaccid vesicle membrane: Wadh= τ_m (1−cos θc)

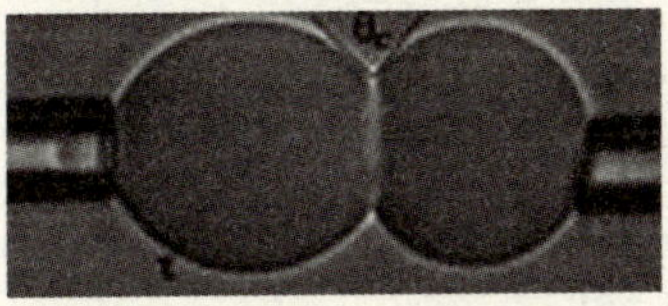

Figure 3: two vesicles in contact

Adhesion energy values for the controls and for the Le^x / Le^x experiments are given in Table 1. The effect of adding calcium on the Le^x vesicles is clearly seen in the table and the adhesion energy in $CaCl_2$ is 2.5 times higher than in NaCl, whereas the Le^x / Lac and Lac/Lac experiments showed a small decrease of the adhesion energy in calcium (3.10^{-6} J/m^2). For comparison, calcium has no influence on the adhesion of pure SOPC vesicles. These results show unambiguously that Le^x groups are necessary on both vesicles for the calcium-induced adhesion enhancement to occur. This is in agreement with the specific interaction scheme advocated by Hakomori.

Table 1 Adhesion energy（$10^{-6}J/m^2$）

left vesicle / right vesicle	in NaCl 0.2M	in $CaCl_2$ 0.11M
Le^x / Le^x	4.5 ± 2	11 ± 2
Le^x / Lac	5.4 ± 1	2.5 ± 2
Lac/ Lac	9.5 ± 0.5	6 ± 1
pure SOPC vesicles	14 ± 2	15 ± 4

This first direct measurement of biologically relevant ultra-weak carbohydrate-carbohydrate interaction shows that it is possible to quantify them even if they are smaller than the thermal energy，and opens up a new promising field in biology.

2.2 *Direct measurements of Le^x-Le^x interaction using a natural Le^x glycolipid.*

By using a vesicle micromanipulation approach with chemically synthesized natural Le^x pentasaccharide glycosphingolipid，we have demonstrated that in contrast to glyconeolipids which allow strong orientational freedom of the Le^x group，the natural lipid showed a restricted orientation of the Le^x group. The adhesion induced by Le^x-Le^x interaction was thereby considerably enhanced，indicating that relative orientation of the two Le^x groups is a predominant factor in Le^x-Le^x recognition.

2.3 *Demonstration of homotypic characteristic of Le^x-Le^x interaction using an isomer of Le^x glycolipid.*

In another experiment we replaced the Le^x trisaccharide determinant in the head group by its isomer Lewis^a（Le^a）trisaccharide，in which the galactose and fucose are permutated relative to Le^x on one vesicle surface. The adhesion energy observed for Le^x-Le^a pair was very weak, confirming the homotypic characteristic of this type of carbohydrate-carbohydrate interactions.

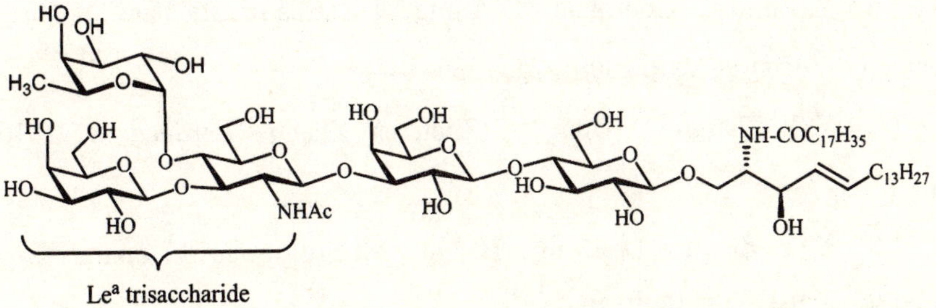

Le^a glycolipid

3. Perspective

Weak biological processes have been discovered only recently, and the number of established carbohydrate-carbohydrate interactions is smaller than that of the known protein-protein interactions. The study of carbohydrate-carbohydrate interactions has the potential to become a hot topic of glycobiology in the near future.

References

[1] I. Eggens, B. A. Fenderson, T. Toyokuni, B. Dean, M. R. Stroud, S. Hakomori, *J. Biol. Chem.* 1989, 264, 9476–9484.

[2] N. Kojima, B. A. Fenderson, M. R. Stroud, R. I. Goldberg, R. Habermann, T. Toyokuni, S. Hakomori, *Glycoconjugate J.* 1994, 11, 238–248.

[3] F. Pincet, T. Le Bouar, Y. Zhang, J.–M. Mallet, E. Perez, P. Sinaÿ, *Biophys. J.* 2001, 80, 1354–1358.

[4] C. Gourier, F. Pincet, E. Perez, Y. Zhang, J.–M. Mallet, P. Sinaÿ, *Glycoconjugate J.* 2004, 21, 165–174.

[5] C. Gourier, F. Pincet, E. Perez, Y. Zhang, Z. Zhu, J.-M. Mallet, P. Sinaÿ, *Angew. Chem. Int. Ed.* 2005, 44, 1683-1687.

[6] Y. Luo, C. Gourmala, D. Dong, F. Barbault, B. Fan, Y. Hu, Y. Zhang. *Glycoconjugate J.* 2008, 25, 335-344.

[7] Y. Y. Zhang, D. Dong, T. Zhou, Y. Zhang, *Tetrahedron*, 2010, 66, 7373-7383.

[8] Y. Y. Zhang, D. Dong, H. Qu, M. Sollogoub, Y. Zhang. *Eur. J. Org. Chem.* 2011, 7133-7139.

作者简介

Dr. Yongmin Zhang graduated from the Department of Pharmaceutical Sciences of Beijing Medical College （current School of Pharmaceutical Sciences，Peking University）in January 1982. He got his Ph.D. of Medicinal Chemistry at South-Paris University （University of Paris 11）in July 1986. After two years post doctoral research at Beijing Medical University （current Peking University Health Science Centre），he worked as a visiting associate professor then a visiting professor at South-Paris University prior to taking a permanent position at CNRS （French National Scientific Research Centre）in 1991，where he is now the director of research at the CNRS-Sorbonne University joint research institute. He is a laureate of several research awards，such as "Prize of medicinal chemistry research" which he got from the 25th International Meeting of Medicinal Chemistry. He was elected corresponding member of the French National Academy of Pharmacy in 2008，then became a member of the French National Academy of Pharmacy in 2012. Inventor of 4 patents，Dr. Yongmin Zhang has published more than 270 original research papers. His current research interests include biologically active oligosaccharides，glycosphingolipids，carbasugars，azasugars，modified cyclodextrins，and fullerene chemistry，aiming at the carbohydrate-based drug development.

Overview of the Plant Chromatin and Epigenetics Research Field

Daoxiu Zhou

In eukaryotes, the genomic DNA is tightly compacted into a complex structure known as chromatin. To control genome activities, the accessibility of chromatin is dynamically regulated during growth and development of the organisms, which constitutes the basis of epigenetic regulation. Plants are sessile organisms that have to adapt to the changing environment. It is therefore essential for plants to develop strategies to be able to respond rapidly and flexibly to environmental changes for their adaptation and survival, which requires rapid and reversible changes of chromatin structure permitting emergence of specific responsive gene expression profiles. During the last years, considerable progress has been made in the study of chromatin basis of epigenetic regulation in plants, which has revealed not only novel epigenetics mechanisms that are unique to plants, but also open wide perspectives for the future study.

Plant epigenomics:

Chromatin structure is regulated by an array of proteins or protein complexes, leading to specific profiles of chromatin modification and remodelling, known as Epigenomes, which include genome-wide DNA methylation, covalent modification of the N-terminal tails of core histones, histone variant deposition, nucleosome positioning and compaction and chromatin protein association, *etc.* Genome-wide profiles of DNA methylation, histone modifications and histone variant deposition have been determined in different organ/developing stages of Arabidopsis and a few other plant species [1-5]. Those studies not only reveal characteristic association of specific chromatin modification within distinct genomic domains, but also show agonistic and antagonistic relationship between different modifications. Association of the epigenomic marks with gene expression levels allows the identification of specific chromatin signatures related to gene activity [6]. However, except pollen cells, no cell type-specific epigenome is known in plants. In addition, it remains largely unknown how are regulated epigenome emergence, maintenance and resetting during cell division/differentiation and plant development process.

Chromatin modifications and recognition mechanism:

During the last years, enzymes involved in DNA methylation, histone modification and chromatin remodeling have been identified and studied in plants [7-8]. DNA methylation plays a primary role in repetitive sequence silencing and heterochromatin formation in plants. Plant DNA methylation system displays a number of specific feature compared to other eukaryotes. In particular the siRNA-dependent DNA methylation pathway implicates specific protein complexes such as the RNA polymerase VI and V that are unique to plants [7]. Covalent modifications of the N-terminal tails of the core histones play pivotal roles in chromatin remodelling and in gene regulation. Histone modifications include acetylation, methylation, phosphorylation, ubiquitination, among others. Most of the modifications are on the lysine, arginine or serine residues of the core histones. Specific patterns of histone modifications determine active or repressed states of the associated DNA. Plant histone modification and recognition systems show also a number of specific features. For instance histone H3K9me2 is prevalently associated with heterochromatin and H3K9me3 is associated with expressed genes, whereas H3K9me2/3 are both involved in gene silencing in animal cells [8]. The Heterochromatin Protein 1 （HP1）binds to H3K9me2/3 and induces heterochromatin formation, while the plant homolog called LHP1 recognizes H3K27me3. Although a number of protein motifs involved in binding to methylated DNA and histones have been studied, the recognition mechanisms of chromatin modification marks are not fully understood. How histone modifications cross-talk among them and with DNA methylation remains also to be studied, which may requires elucidation of functional relationship between different chromatin regulators.

Epigenetic regulation of plant development:

Plant life cycle is characterized by distinct developmental phases. Reprogramming of gene expression must occur during the transitions of plant development and depends on chromatin modification and remodeling. The role of DNA methylation and histone modification involved in several classic epigenetic phenomena affecting plant development and growth such as gene imprinting, vernalization, paramutation and gene silencing are being elucidated at very advanced levels [7, 8]. Many of the chromatin regulators have been shown to play an important role in the plant development transition and gene expression reprogramming. For instance the Polycomb group （PcG） Repressive complexes 2 （PRC2）involved in the methylation of histone H3 lysine 27 is shown to play an essential role in embryo/seed development, meristem activity, flowering time and leaf/floral organogenesis [9]. In addition, involvement of chromatin regulators in biotic and abiotic stress-responses of plants has been largely demonstrated in recent years [10, 11]. Cell or tissue-specific gene expression profiles should be maintained across cell divisions, but they should be reset during developmental phase transitions by epigenetic processes which are, however, not clearly understood at the present time. It also remains to know how locus-specificity of chromatin factors in the genome is determined.

Epigenetic inheritance of adaptive responses:

Natural epigenetic variation exists between populations or among individuals with a similar genotype. For instance, DNA methylation in genes is extremely polymorphic among natural accessions of Arabidopsis [12]. Recent studies by examining spontaneously occurring variation in DNA methylation in Arabidopsis plants propagated by single seed descent for 30 generations indicate that transgenerationally stable changes in cytosine methylation occur at a relatively high frequency, suggesting that epigenetic variation in plant populations exceeds genetic diversity and that it is influenced by the environment [13, 14]. Therefore, not only do epigenetic phenomena modulate the activity of the genome in response to environmental stimuli, but they also constitute a potential source of natural variation. Variation in DNA methylation have contributed to phenotypic variation in plants, as a number of epimutations (epialleles) affecting plant development or important agronomic traits have been identified [15]. Recent work has suggested that abiotic and biotic stresses including DNA damage, drought, high salinity and pathogens are sources of epigenetic variation [15]. In addition, a number of studies have revealed that epigenetic variability in plants can be enhanced by various stress treatments [16]. Some of the epigenetic adaptive responses to environmental cues can be transmitted to next generations [16, 17]. Understanding the emergence and heritability of epigenetic variation is critical for understanding how they might become subject to natural selection and thus affect genetic diversity. For the long-term transgenerational adaptation to environmental cues, the perceived information must be memorized in an epigenetic form that is propagated through mitotic and meiotic divisions. Multiple epigenetic mechanisms （*e.g.* DNA methylation，histone modification，non-coding RNA，*etc.*）have been suggested to stabilize and buffer the epigenetic states of gene expression. It will be of primary importance to answer how adaptive epigenetic information is transmitted to next generations or otherwise reset，which are not clear at the present time.

References

[1] Zhang X，Yazaki J，Sundaresan A，Cokus S，Chan SW，Chen H，Henderson IR，Shinn P，Pellegrini M，Jacobsen SE，*et al.*：*Cell* 2006，126：1189–1201.

[2] Zemach A，McDaniel IE，Silva P，Zilberman D：*Science* 2010，328：916–919.

[3] Zhang X，Bernatavichute YV，Cokus S，Pellegrini M，Jacobsen SE：*Genome Biol* 2009，10：R62.

[4] Wollmann H，Holec S，Alden K，Clarke ND，Jacques PE and Berger F：*Plos Genet* 2012，8：e1002658.

[5] Zilberman D，Coleman-Derr D，Ballinger T，Henikoff S：*Nature.* 2008 Nov 6；456（7218）：125–129

[6] Roudier F，Ahmed I，Berard C，Sarazin A，Mary-Huard T，Cortijo S，Bouyer D，Caillieux E，Duvernois-Berthet E，Al-Shikhley L，*et al.*：*Embo J* 2011，30：1928–1938.

[7] Gehring，M. and Henikoff，S：Biochim. Biophys. Acta 2007，1769：276–286. Haag，J.R.，and Pikaard，C.S：*Nat. Rev. Mol. Cell Biol.*

2011, 12: 483–492.

[8] Berr A., Shafiq S, Shen WH: Biochim Biophys Acta 2011, 1809 (10): 567–576.

[9] Hennig L, Derkacheva M: *Trends Genet.* 2009, 25 (9): 414–23.

[10] Servet C., Conde e Silva N, Zhou DX.: *Mol Plant.* 2010, 3: 670–677.

[11] Chen ZJ and Tian L: *Biochim Biophys Acta* 2007, 1769: 295–307.

[12] Vaughn MW, *et al.*: *Plos biology* 2007, 5 (7): e174.

[13] Becker C, *et al. Nature* 2011, 480 (7376): 245–249.

[14] Schmitz RJ, *et al.*: *Science* 2011, 334 (6054): 369–373.

[15] Paszkowski J, Grossniklaus U: *Curr Opin Plant Biol.* 2011, 14 (2): 195–203.

[16] Gutzat R, Mittelsten Scheid O: *Curr Opin Plant Biol* 2012.

[17] Boyko A, Blevins T, Yao Y, Golubov A, Bilichak A, Ilnytskyy Y, Hollunder J, Meins F, Jr., Kovalchuk I: *Plos One* 2010, 5: e9514.

[18] Feschotte C, Jiang N, Wessler SR: *Nat Rev Genet* 2002, 3: 329–341.

[19] P*ontier D, Picart C, Roudier F, Garcia D, Lahmy S, Azevedo J, Alart E, Laudié M, Karlowski WM, Cooke R, Colot V, Voinnet O, Lagrange T*: Mol Cell. 2012, 48 (1): 121–32.

作者简介

Daoxiu Zhou did his PhD thesis at the Université de Grenoble on chloroplast biology (1984—1988), postdoctoral research at the University of California, San Francisco (1988—1991) on gene transcriptional regulation. In 1990 he was recruited by CNRS as Chargé de recherche (1991—1996) to study plant transcription factors. In 1996 he became a professor at the University of Amiens (1996—2000) and continued the research on plant transcription factors. In 2000, he was appointed to the University of Paris 11 at Orsay where he started to work in the field of plant chromatin and epigenetics. At Université Paris 11, he leads a research team working on the chromatin regulation of gene expression in Arabidopsis, a model plant species. He has been also a visiting professor at the Huazhong agricultural University where he directs a research team on rice epigenetics.

During the last decade, his team has been concentrated on the study of histone modification and recognition mechanism in Arabidopsis and rice. He has studied the regulatory mechanisms of several histone acetyltransferase (HAT), deacetylase (HDAC) and histone demethylases both in Arabidopsis and rice. Most significantly, the work has revealed that a GCN5, a major HAT in Arabidopsis, is required for acetylation of histone H3 lysines 9 and 14 and targets to a large number of genes genome-wide. Work on HDAC has revealed an important role of histone deacetylation in epigenetic interaction between parental genomes to control non-additive gene expression in hybrid rice. A model for molecular mechanism of histone acetylation/deacetylation-mediated activation of short-term regulated genes has been proposed. Recently his team has evaluated the role of histone methylation and histone variant H2A.Z in rapid activation. Our data show that changes of histone H3K4me3 and H3K27me3 levels lagged behind gene activation, suggesting that these methylation marks may play a role to label gene activity rather than to activate/ repress gene transcription. Recently we have demonstrated that a CHD3 protein could recognize and regulate H3K4 and H3K27 methylation over under-expressed genes in the rice genome, revealing a reading mechanism of a combination of histone modification marks by chromatin proteins. In addition, function of WUSCHEL-related homeobox (WOX) genes in controlling plant development has been studied in both Arabidopsis and rice.

Currently his research is focused on the characterization of chromatin regulators for their function in epigenomic dynamics, in recognition and interpretation of epigenetic marks and in gene regulation and plant development control. Special efforts are dedicated to study the function of chromatin regulators involved in emergence, maintenance/resetting and inheritance of chromatin-based epigenetic information in rice aiming to search for strategies to alter the stability and inheritance of adaptive epigenetic information for rice genetic improvement.

Ten selected recent publications

[1] Li T, Chen X, Zhong X, Zhao Y, Liu X, Zhou S, Cheng S, Zhou DX (2013). Histone demethylase JMJ705-mediated remove of histone H3 lysine 27 trimethylation is involved in defense-related gene activation in rice. ***Plant Cell***; 25 (11): 4725–4736

[2] Chen Q, Chen X, Wang Q, Zhang F, Lou Z, Zhang Q, Zhou DX (2013). Structural basis of a histone H3 lysine 4 demethylase required for stem elongation in rice. ***PLoS Genet.*** 9 (1): e1003239.

[3] Chen X, Zhou DX (2013) Rice epigenomics and epigenetics: challenges and opportunities. ***Curr Opin Plant Biol.*** 16 (2): 164–169.

[4] Hu Y, Liu D, Zhong X, Zhang C, Zhang Q, Zhou DX (2012) A CHD3 protein recognizes and regulates methylated histone H3 lysines 4 and 27

over a subset of targets in the rice genome. ***Proc. Natl. Acad. Sci. USA*** 109（15）: 5773–5778.

[5] Zhao Y, Hu Y, Dai M, Huang L, Zhou DX （2009）. The Wuschel-related homoeobox gene WOX11 is required to activate shoot-borne crown root development in rice. ***Plant Cell***, 21: 736–748.

[6] Kim, W. Benhamed, M. Servet, C. Latrasse, D. Zhang, W. Delarue, D. Zhou DX （2009）Histone acetyltransferase GCN5 interferes with the miRNA pathway in Arabidopsis. ***Cell Res***. 19（7）: 899–909.

[7] Sun Q & Zhou DX （2008）. Rice jmjC domain-containing gene ***JMJ 706*** encodes H3K9 demethylase required for floral organ development. ***Proc. Natl. Acad. Sci. USA***, 105: 13679–13684.

[8] Huang L, Sun Q, Qin F, Li C, Zhao Y, Zhou DX （2007）. Down-regulation of a Silent Information Regulator2-related gene induces DNA fragmentation and cell death in rice. ***Plant Physiol***, 144: 1508–1519.

[9] Dai M, Hu Y, Zhou Y, Liu H, Zhou DX （2007）. A rice ***WUSCHEL-LIKE HOMEOBOX***（***WOX***）gene represses a YABBY gene expression required for shoot development. ***Plant Physiol***, 144: 380–390.

[10] Dai M, Zhao Y, Ma Q, Hu Y, Heden P, Zhang Q, Zhou DX （2007）. The Rice YAB1 transcription factor gene is involved in the feedback regulation of gibberellin metabolism. ***Plant Physiol***, 144: 121–133.

Density Depletions and Energetic Electrons Associated with Magnetic Reconnection Observed by Cluster

Shiyong Huang（黄狮勇）

Laboratoire de Physique des Plasmas, CNRS-UPMC-Ecole Polytechnique, Palaiseau, 91128, France

Abstract: Density depletions and energetic electrons were simultaneity observed associated with magnetic reconnection in the magnetotail by Cluster spacecraft. Intense electric field, parallel current with respect to the ambient magnetic field, and strong wave activities were observed in the density depletion layer, while weak electric field and weak wave activities, and antiparallel current were detected in the vicinity of X-line. Energetic electron fluxes increase in the density depletion layers around separatrix region, indicating the electrons are accelerated there. In addition, there were electron beams antiparallel to the magnetic field with flat-top distribution in the density depletion layer, and parallel electron beams in the vicinity of X-line. We discussed the density depletions, energetic electrons and electron beams in the two distinct regions, and proposed electron acceleration mechanism.

Keywords: Density Depletion; Energetic Electron; Magnetic Reconnection; Electron Acceleration.

1. Introduction

Magnetic reconnection is a fundamental physical process in laboratory and astrophysical plasmas, which converts magnetic energy into plasma kinetic and thermal energy, and enables reconfiguration of the magnetic field topology. Magnetic separatrix is a boundary layer between inflow and outflow region, and play an important role in the magnetic reconnection process. Density depletion layer, associated with strong Hall currents and high magnetic pressure, lies just down stream of the magnetic separatrix [22]. Hall MHD simulations have confirmed the essential role of Hall terms in the formation of density depletion layers near the separatrices [25]. Recently, *Huang et al.* [11] have found the density depletion is outside of the peak of the out-of-plane magnetic field via particle in cell simulation. The structures of density depletion have also been studied by satellite observations. *Øieroset et al.* [2001] observed a density dip in the diffusion region in the distant tail. *Retinò et al.* [2006] reported density depletion layer in the magnetopause separatrix region, and found that it is associated with strong electric field, electron beams, and wave activities. *Khotyaintsev et al.* [2006] detected a density cavity which is formed due to the escape of magnetospheric electrons along the newly opened field lines.

Energetic electrons have been widely investigated by simulations [*Drake et al.*, 2005; *Fu et al.*, 2006; *Hoshino et al.*, 2001a, 2001b; *Pritchett et al.*, 2006] and observations [*Chen et al.*, 2008; *Wang et al.*, 2008; *Retinò et al.*, 2008; Huang et al., 2012a, 2012b] during magnetic reconnection. Pritchett et al. [2006] found the parallel electric field that exists in the low-density cavities along two of the separatrices leads to a cold electron beam during guide field magnetic reconnection. This beam is funneled into the near vicinity of the X-line where the electrons are further accelerated by the parallel electric field. *Wang et al.* [2008] have observed spatial distribution of energetic electrons during magnetic reconnection, which is consistent with the simulation of *Fu et al.* [2006].

However, until now what role the density depletion plays, and what relationship between density depletion and energetic electrons in the magnetic reconnection process is still an open question via observations.

In this paper, we observe density depletions and energetic electrons simultaneity by the Cluster in the near-Earth magnetotail and compare with the previous simulations.

We use the data from several different instruments on board Cluster. Magnetic field measurements are obtained from the fluxgate magnetometer (FGM) [*Balogh et al.*, 2001], and the ion plasma data are taken from the Cluster Ion spectrometer (CIS) experiment [*Rème et al.*, 2001]. The high energy electrons data are obtained from the research with adaptive particle imaging detectors (RAPID) experiment [*Wilken et al.*, 2001]. The low to middle energy electron data with 4s time resolution are from the Plasma Electron and Current Experiment (PEACE) instrument [*Johnstone et al.*, 1997]. The high frequency wave Electric and magnetic field spectrograms data are from Spatio-Teporal Analysis of Field Fluctuations (STAFF) experiment [*Cornilleau-Wehrlin et al.*, 2003]. The electric field and spacecraft potential are measured by the electric field and waves (EFW) instrument [*Gustafsson et al.*, 2001].

2. Observation overview

Figure 1 shows the observations of the magnetic reconnection event on 15 September 2001 from 05: 00 UT to 05: 06 UT. All the variables are presented in the GSM (Geocentric Solar Magnetospheric) coordinate except the electric field, which is shown in the spin plane. The Cluster located in the near-Earth magnetotail, approximately –18.6 R_E (the radius of Earth) from the centre of the Earth during this time period. This event was first examined by *Xiao et al.* [2007], who estimated the fast reconnection rate which falls in the range predicted by steady reconnection simulations.

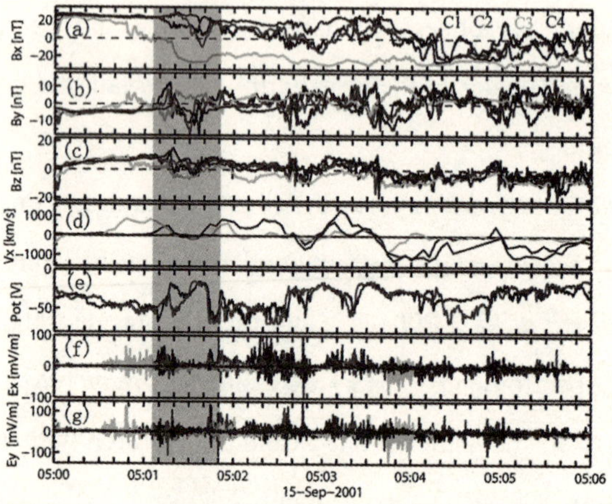

Figure 1　Observation overview of four Cluster.
(a) – (c) three components of magnetic field, (d) the X component of ion plasma flow, (e) potential, and (f) – (g) two components of electric field.

During the interval of 05: 00: 30–05: 02: 10 UT, Cluster 3 firstly detected earthward flow (positive Vx), up to 1000km/s, then Cluster 1 also detected earthward flow around 05: 01: 00 UT. Latterly, Cluster 1 and Cluster 3 observed small tailward flow (around 05: 01: 20 UT), but soon they were back to the earthward flow. At the same time, the magnetic field Bz change from the positive values to the negative values, then to positive values. The reversals of Vx and Bz indicates that the Cluster crossed from the earthward side of an active diffusion region to the tailward side, then back to the earthward side. From 05: 02: 10 UT to 05: 06: 00 UT, the similar crossing can be identified. *Xiao et al.* [2007] have identified Hall effects and confirmed the crossing diffusion region in details. This event has not obvious guide field. The subinterval of our interesting time, i.e. 05: 00: 50 ~ 05: 02: 10 UT, was been signed in Figure 1 with pink shaded, when density depletion and energetic electrons were observed simultaneity.

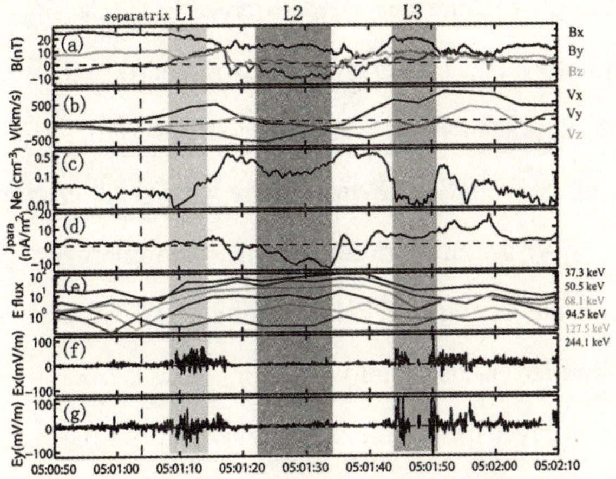

Figure 2 Detailed observations of density depletions observed by Cluster 1. From top to bottom panel: magnetic field, ion plasma flow, plasma density, current density parallel to the ambient magnetic field, energetic electron fluxes, x and y component of electric field in GSE coordinate.

The spacecraft potential can be measured in all four Cluster satellite, but these data of Cluster 3 and 4 were some contamination. Moreover, there is no plasma data for Cluster 2. Therefore, Figure 2 only shows the observations of Cluster 1. The electron density (Figure 2c) is derived from the spacecraft potential measurements [*Pedersen et al.*, 2008]. We divide the whole interval into three subintervals: L1, L2, and L3 according to the location of

Cluster 1 in the diffusion region. Cluster 1 located in the northward of center plasma sheet with positive Bx. The Hall signatures are detected by Cluster 1 during all three subintervals, which implies the Cluster spacecraft crossed ion diffusion region. During the subinterval L1, Cluster 1 stayed in the earthward flow （$Vx>0$）, i.e. earthward of X-line, and detected density depletion with a dip of plasma density （Figure 2c）. Intense electric field, up to 100 mV/m （Figure 2f and 2g）, parallel current （Figure 2d）and energetic electron fluxes enhancement （Figure 2e）were observed in the density depletion layer. During the subinterval L2, tailward flow was small, only up to-200km/s, *Bx* is not large, less than 15 nT, and *Bz* is almost zero, which suggests the Cluster 1 was in the vicinity of X-line. There were small electric field, which is less than 25 mV/m, antiparallel current, and energetic electron fluxes enhancement in the vicinity of X-line. In addition, a small dip of plasma density was observed in the vicinity of X-line. The Cluster 1 was back to the density depletion layer during the subinterval L3. Similarly, Cluster 1 also observed intense electric field and parallel current, and energetic electron fluxes enhancement. The density depletion （05：01：09 UT）with *Bx*–22nT is outside of the peak of the out-of-plane magnetic field （05：01：17 UT） with *By*–12nT, which is in accordance with the simulations ［*Huang et al.*, 2008］.

Figure 3 displays power spectrum density of magnetic and electric field. Intense wave activities of magnetic electric field around the lower hybrid frequency were detected in the density depletion layers （L1 and L3）. However, the wave activity became weak in the vicinity of X-line （L2）.

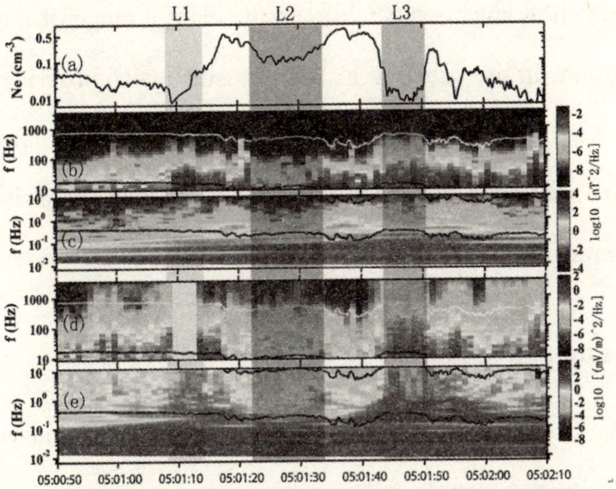

Figure 3 The wave spectrogram of magnetic and field. (a) the plasma density, (b) - (c) the high and low frequency wave spectrogram of the magnetic field from STAFF and FGM instruments, respectively. (d) - (e) the high and low frequency wave spectrogram of the electric field from STAFF and EFW instruments, respectively. The yellow line, black line, and blue line are the electron cyclotron frequency, the lower hybrid frequency, and the ion cyclotron frequency, respectively.

The evolution of electron spectra as Cluster 1 crossed the density depletion regions and vicinity of X-line is shown in Figure 4. Figure 4a-b, c-d, e-f correspond to the subinterval L1, L2, and L3 respectively. At the subinterval L1, Cluster 1 staying in the density depletion layer, observed electron beam (~2 keV) antiparallel to the magnetic field, which covers low and high energy components. The electron distributions had flat-top feature: the PSD was almost constant between 180eV and 1.5keV but sharply decreased at 3 keV. The flat-top distribution was also observed in the vicinity of X-line during the subinterval L2, but the PSDs of parallel direction are higher than two other directions. The PSDs of different energy range in the vicinity of X-line were larger than ones in the density deletion layer，specially，in the energy range of the shoulder of flat-top distribution. In the subinterval L3，Cluster 1 also observed antiparallel electron beam with slightly flat-top distribution. In addition，parallel electron beam were also observed in the density depletions，which is consistent with the crossing of separatrix ［*Retinò et al.*，2006］.

3. Discussion and summary

We observe density depletion layers in the diffusion region without guide field by Cluster. Intense electric field，strong wave activities around the lower hybrid frequency were detected in the density depletion layer. Moreover，Cluster crossed the vicinity of X-line with weak electric field and wave activities.

The electron beam antiparallel to the magnetic field，as well as current parallel to the magnetic field was observed in the density depletion layer. It suggests that the current was mainly generated by the electron beam.

The flat-top distribution has been reported in many literatures ［*Asano et al.*，2008；*Chen et al.*，2009；*Wang et al.*，2010］. In our event，we observed the flat-top distribution in the density depletion layer and vicinity of X-line. In addition，the flat-top distribution is always accompanied by the electron beam，which is in accordance with the predication by *Hoshino et al.* ［2001b］，who suggest that the flat-top distribution is the thermalized component of the electron beam toward the X-line.

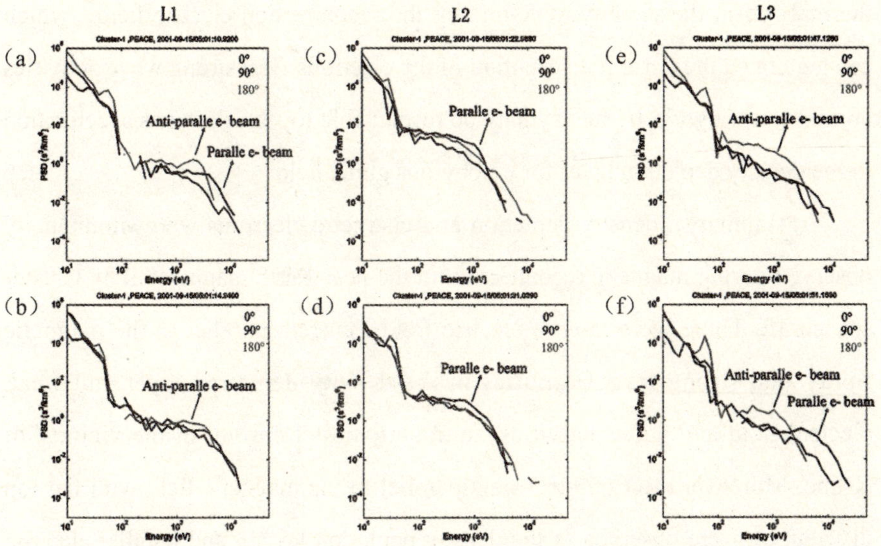

Figure 4 Electron spectra in the three subintervals measured by Cluster 1. The L1 (a–b), L2 (c–d) and L3 (e–f) periods correspond to the three subintervals shown in Figure 2.

The most interesting thing in our event is the energetic electron fluxes enhancement. *Pritchett et al.* [2006] found that cold electron beam in the low-density cavities is funneled into the near vicinity of the X-line where the electrons are further accelerated by the parallel electric field during the magnetic reconnection with guide field. In our event, the energetic electron fluxes gradually enhanced when Cluster toward the X-line, and low energy electron beam antiparallel to the magnetic field was observed in the density depletion layer. However, the largest energetic fluxes were in the vicinity of X-line during the subinterval L2, which suggests the electrons were further accelerated in this region. This observation is in contrast to *Fu et al.* [2006]'s simulation. Therefore, we proposed that: without obvious guide field, the low energy electrons were firstly accelerated, flowed toward the X-line and funneled into the near vicinity of X-line, then these electrons were further accelerated and thermalized in the vicinity of X-line by the reconnection electric field，which can generated the flat-top distribution of the electrons. The strong wave activities around lower hybrid frequency may be responsible for the electrons acceleration in the density depletion layer for no obvious guide field.

In summary，density depletion and energetic electrons were simultaneity observed during magnetic reconnection in the near-Earth magnetotail by Cluster spacecraft. There were intense electric field，current parallel to the magnetic field，and strong wave activities in the density depletion layer and weak electric field and wave activities，and antiparallel current in the vicinity of X-line. Moreover electron beams antiparallel to the magnetic field with flat-top distribution were observed in the density depletion layer，and parallel electron with flat-top distribution were detected in the vicinity of X-line. We discussed the density depletion，energetic electrons and electron beam in the two regions，and proposed the electron beams were accelerated by wave activities around lower hybrid frequency and flowed toward the X-line in the density depletion layer，then these electrons were further accelerated in the vicinity of X-line during the no obvious guide field magnetic reconnection.

References

[1] Asano，Y.，*et al.* Electron flat-top distributions around the magnetic reconnection region，*J. Geophys. Res.*，2008，113，A01207，doi: 10.1029/2007JA012461.

[2] Balogh，A.，*et al.* The Cluster magnetic field investigation: overview of in-flight performance and initial results，*Ann. Geophys.*，2001，19，1207–1217.

[3] Cornilleau-Wehrlin，N.，et al.：First results obtained by the Cluster

STAFF experiment, Ann Geophys, 21, 437–456, 2003

[4] Chen, L.-J., *et al*. Observation of energetic electrons within magnetic islands, *Nature Phys.*, 2008, 4, 19– 23, doi: 10.1038/nphys777.

[5] Chen, L. J. *et al*. Multispacecraft observations of the electron current sheet, neighboring magnetic islands, and electron acceleration during magnetotail reconnection, *Phys. Plasmas*, 2009, 16, 056501.

[6] Drake, J. F., M. A. Shay, W. Thongthai, and M. Swisdak, Production of energetic electrons during magnetic reconnection, Phys. Rev. Lett., 2005, 94, doi: 10.1103/PhysRevLett.94.095001.

[7] Gustafsson, G., *et al*.: First results of electric field and density observations by CLUSTER EFW based on initial months of operation, Ann. *Geophys*., 2001, 19, 1219–1240.

[8] Fu, X. R., Q. M. Lu, and S. Wang The process of electron acceleration during collisionless magnetic reconnection, *Phys. Plasmas*, 2006, 13, 012309.

[9] Hoshino, H., T. Mukai, T. Terasawa, and I. Shinohara Suprathermal electron acceleration in magnetic reconnection, *J. Geophys. Res.*, 2001a, 106 (A11), 25979−25997, doi: 10.1029/2001JA90005

[10] Hoshino, M., K. Hiraide, and T. Mukai, Strong electron heating and non-Maxwellian behavior in magnetic reconnection, *Earth Planets Space*, 2001b, 53, 627–634.

[11] Huang C, Wang R S, Lu Q M, *et al*. Electron density hole and quadruple structure of By during collisionless magnetic reconnection. Chinese Sci Bull, doi: 10.1007/s11434−009−0538−z

[12] Huang, S. Y., A. Vaivads, Y. V. Khotyaintsev, M. Zhou, H. Fu, A. Retinò, X. H. Deng, M. André, C. M. Cully, J. He, F. Sahraoui,

Z. Yuan, and Y. Pang, Electron Acceleration in the Reconnection Diffusion Region: Cluster Observations, *Geophys. Res. Lett.*, 2012a, doi: 10.1029/2012GL051946

[13] Huang S. Y., X. H. Deng, M. Zhou, Z. G. Yuan, H. M. Li, and D. D. Wang, Energetic electrons associated with magnetic reconnection in the sheath of interplanetary coronal mass ejection, *Chin Sci Bull*, 2012b, doi: 10.1007/s11434-012-4974-9

[14] Johnstone, A. D., *et al.*: PEACE: A plasma electron and current experiment, Space Sci. Rev., 79, 351–398, 1997.

[15] Khotyaintsev, Yu., Vaivads, V.A., Retinò, A., André, M., Owen, C.J. and Nilsson, H.: Formation of the inner structure of a reconnection separatrix region, *Phys. Rev. Lett.*, 2006, 97, 205003, doi: 10.1103/PhysRevLett.97.205003.

[16] Pedersen, A., *et al.*, Electron density estimations derived from spacecraft potential measurements on Cluster in tenuous plasma regions, *J. Geophys. Res.*, 2008, 113, A07s33, doi: 10.1029/2007JA012636.

[17] Pritchett, P. L., Relativistic electron production during guide field magnetic reconnection, *J. Geophys. Res.*, 2006, 111, A10212, doi: 10.1029/2006JA011793.

[18] Øieroset M, Phan T D, Fujimoto M, et al. In situ detection of collisionless reconnection in the Earth's magnetotail. Nature, 2001, 412: 414-417.

[19] Retinò, A., *et al.*: Structure of the separatrix region close to a magnetic reconnection X-line: Cluster observations, *Geophys. Res. Lett.*, 2006, 33, L06101, doi: 10.1029/2005GL024650.

[20] Retinò, A., *et al.*, Cluster observations of energetic electrons

and electromagnetic fields within a reconnecting thin current sheet in the Earth's magnetotail, J. Geophys. Res., 2008, 113, A12215, doi: 10.1029/2008JA013511.

[21] Rème, H., *et al.*, First multispacecraft ion measurements in and near the Earth's magnetosphere with the identical Cluster ion spectrometry (CIS) experiment, *Ann. Geophys.*, 2001, 19, 1303.

[22] Shay, M. A., *et al.*: Alfvenic magnetic field reconnection and the Hall term, *J. Geophys Res.*, 2001, 106, 3759.

[23] Wilken, B., *et al.*, First results from the RAPID imaging energetic particle spectrometer on board Cluster, *Ann. Geophys.*, 2001, 19, 1355–1366.

[24] Xiao, C. J., *et al.*, Cluster measurements of fast magnetic reconnection in the magnetotail, Geophys. Res. Lett., 2007, 34, L01101, doi: 10.1029/ 2006GL028006.

[25] Yang, H. A., Jin, S.P. and Zhou, G.C.: Density depletion and Hall effect in magnetic reconnection, *J. Geophys. Res.*, 2006, 111, A11223, doi: 10.1029/2005JA011536.

作者简介

黄狮勇（Shiyong Huang）　Email: shiyonghuang@msn.com

教育经历：

2008.09—2013.06　武汉大学，电子信息学院，硕博连读，理学博士

2012.10—2012.11　北京大学，访问研究

2011.09—2012.09　瑞典，空间物理研究所，国家公派联合培养

2011.05—2011.06　法国，等离子体物理实验室，访问研究

2004.09—2008.06　武汉大学，物理科学与技术学院，理学学士

工作经历：

2019/04—至今　武汉大学，电子信息学院，教授

2016/03—2019/03　武汉大学，电子信息学院，特聘研究员

2014/02—2015/09　法国，巴黎第六大学，Post-doc research fellow

2013/07—2016/02　武汉大学，电子信息学院，师资博士后

获奖情况：

1) 中组部"万人计划"青年拔尖人才，03/2019

2) 国际无线电联盟青年科学家奖，05/2018

3) 欧洲地球科学学会杰出青年科学家奖，04/2016

4) 欧洲空间局Cluster和双星计划杰出贡献奖，10/2015

5) 湖北省"楚天学子"，03/2015

6) 第八次全国空间天气学研讨会优秀青年论文奖，09/2012

7) 欧洲地球科学年会优秀学生墙报奖，04/2012

8) 武汉大学第七届研究生"十大学术之星"，06/2013

9) 武汉大学研究生学术创新奖一等奖，04/2013

10) 博士研究生国家奖学金，12/2012

研究方向：

1) 与重联相关的电子加速和磁岛/磁通量管

2) 与重联相关的波动和湍流的观测

3) 空间等离子体中电子尺度的湍流和相干结构

4) 磁尾的偶极化锋面

5) 机器学习在空间天气中的应用

主要论文：

已发表论文：>90篇

总引用>1500，h-index: 23.（https://scholar.google.com/citations?user=hXN8A-MAAAAJ&hl=zh-CN）

A Plasma-catalytic Process for Formal dehyde

Min Jia（贾敏）[1]，Zixian Jia（贾子先）[2]

[1]*Qingdao Supply And Marketing Technical Secondary School*，*266200 Qingdao*，*China*（青岛供销职业中等专业学校）

[2]*Laboratoire des Sciences des Procédés et des Matériaux C.N.R.S.*，*Institut Galilée*，*Université Paris* 13，*Sorbonne Paris Cité*，*93430 Villetaneuse*，*France*

1. **Introduction**

The emission of volatile organic compounds（VOCs）by various industrial and agricultural processes is an important source of air pollution and，therefore，a problem for human health and the environment in general[1]. There are many techniques that are studied，developed and industrialized in order to control VOCs emissions. The conventional processes usually consist of trapping（adsorption，filtration，condensation，...）and/or destroying the pollutant（thermal and catalytic oxidation，photocatalysis，....）[2, 3]. The disadvantage of these methods is that they become cost-inefficient and difficult to operate when low concentrations of VOCs need to be treated. With the increased severity of emission limits in mind，this creates the need for an alternative technology that overcomes these weaknesses.

For the abatement of VOCs, non-thermal plasma (NTP) technology has attracted growing interest of scientists over the last 2 decades [4–10]. In NTP, most of energy that is delivered to the system is used to generate high-energy electrons (1–15eV). These high-energy electrons collide with background molecules (N_2, O_2, H_2O, ...) producing secondary electrons, photons, ions and radicals while the background gas remains close to room temperature. These latter species are responsible for the oxidation of VOC molecules, although ionic reactions are also possible. However, the application of NTP for VOCs abatement needs overcome three main weaknesses: incomplete oxidation with emission of harmful compounds (CO, NOx, and other VOCs), a poor energy efficiency and a low mineralization degree.

To solve these problems, researchers are combining the advantages of NTP and catalysis in a technique called plasma–catalysis. The primary idea is that, if the catalysts have a significant adsorption capacity for pollutant molecules, the pollutant retention time can be increased, favoring complete oxidation to CO_2 and H_2O. Interestingly, combining both techniques creates a synergism [11]. Synergetic effects are related to the activation of the catalyst by the plasma. Activation mechanisms include ozone, UV, local heating, changes in work function, activation of lattice oxygen, plasma-induced adsorption/desorption, creation of electron–hole pairs, and direct interaction of gas-phase radicals with adsorbed pollutants.

Various types of catalysts have been used in combination with plasma for VOC decomposition, such as noble metals, transition and group IV and V metal oxides, and zeolites. Among the catalysts, plasmonic nanostructures of noble metals (mainly silver and gold) show significant promise for catalysis [12, 13]. Significantly, it requires the use of metals in a finely divided state, preferably in the form of nanocrystals with precisely controlled properties. The material activity depends on many parameters including composition, size, shape and number density of noble metal nanoparticles. Stable and reproducible morphology remains one of major challenges in the large-scale implementation of plasma catalytic methods. One of the novel approaches in this direction is the photocatalytic growth of the metal nanoparticles on nanoparticulate coatings. Because of the mono-dispersed character of the titanium-oxo-alkoxy（TOA）nanoparticles and their presumed considerable photocatalytic activity, the size of the as grown Ag/Au nanoparticles（NPs）is expected to be governed by Poisson statistics（inherent to the photons absorption process）, which is much narrower compared to the general log-normal one: $\Delta N \approx \sqrt{N}$□N.

In this study, we have developed a plasma-catalytic process of formaldehyde removal using a diphasic process coupling a nano-structured catalyst and an atmospheric pressure plasma. The elaboration of the nanoparticulate catalyst has been firstly studied. Then its performance coupling with plasma has been investigated.

2. Results and discussion

2.1 Catalysts characterization

High efficiency of the traps filling under theband-gap irradiation above 3.2 eV and long-time conservation of trapped electrons have been reported in non-crystalline "oxo" phase. The photocatalytic activity of smallest immobilized non-crystalline titanium oxide oxo-particles has been reported for the ethylene removal. But there is no studies involving the photocatalytic activity of smallest immobilized non-crystalline titanium oxide for the photodeposition application. Hence, first of all, we made it en evidence. The photocatalytic activity of the non-crystalline titania nanoparticles provide the TEM/EFTEM measurements presented in Figure 1.

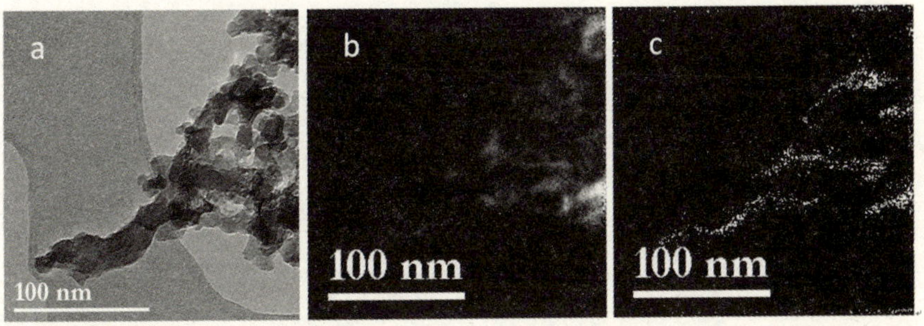

Figure 1 TEM images of silver reduced on the titania nanoparticles immobilized onto ultraporous silica-covered alumina matrix (a) and energy-filtered images of Ti (b) and Ag (c) elemental maps (UV-A irradiation time 5 min).

In the current study, we have used these media in order to validate the photocatalytic silver deposition on nanoscale. Figure 1b evidences a presence of the TOA nanoparticles in the sample, which TEM image is shown in Figure 1a. After 20 min of the sample UV-A irradiation in the silver nitrate solution, the silver deposit can be only seen in the regions occupied by the TOA nanoparticles. Moreover, these images show that the reduced silver atoms do not nucleate a silver nanoparticle at the isolated TOA nanoparticle but seem to be dispersed on its surface.

2.2 Pollutant removal using plasma/silver-catalyst process

In this part, we will present the results obtained in the diphasic process coupling the atmospheric pressure dielectric barrier discharge and a catalytic fluidized bed. The results will be compared with the one obtained in the plasma alone configuration. The influence, in terms of HCHO abatement of the SiO_2

pellets，TiO_2/SiO_2 pellets and $Ag/TiO_2/SiO_2$ pellets will be presented.

The residual acetaldehyde at the reactor exit，as a function of the specific input energy and type of solid particle，is presented in Figure 2.

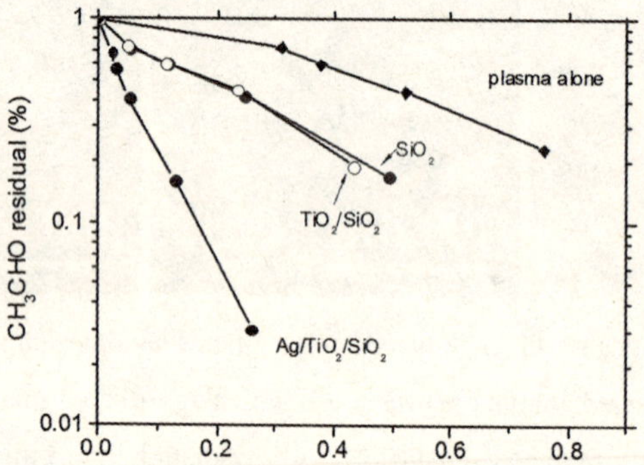

Figure 2　CH_3CHO residual（%）at reactor exit as a function of the specific input energy，the process used and irradiation time.

All CH_3CHO residual values obtained with the fluidized beds（SiO_2，TiO_2/SiO_2 and $Ag/TiO_2/SiO_2$）are lower than those obtained in the plasma alone process. Thus，the decomposition of CH_2CHO can be enhanced by more than a factor of 10，compared to the plasma alone process，depending on the fluidized bed used. For example，at a low power of 0.2W，the residual CH_3CHO reaches 75% in plasma alone process，while the value drops below 3% using Ag /TiO_2/SiO_2 beads in the plasma process.

3. Conclusion

During this Phd work，we have developed a plasma-catalytic process of Acetaldehyde removal using a diphasic process coupling an atmospheric pressure plasma and a nano-structured catalyst. The nanoparticulate catalyst has been firstly studied. Then its performance coupling with plasma has been investigated.

References

[1] Thevenet, F.; Sivachandiran, L.; Guaitella, O.; Barakat, C.; Rousseau, A. Plasma–catalyst coupling for volatile organic compound removal and indoor air treatment: a review. *J. Phys. D. Appl. Phys*. 2014, 47, 224011, doi: 10.1088/0022-3727/47/22/224011.

[2] Jia, Z.; Bouslama, M.; Amar, M. Ben; Amamra, M.; Kayser, M.; Traore, M.; Tieng, S.; Chhor, K.; Chianese, A.; Nadtochenko, V.; Kanaev, A. Nanoparticulate media for environmental applications. *Funct. Mater*. 2013, 20, 417–423, doi: 10.15407/fm20.04.417.

[3] Bouslama, M.; Amamra, M. C.; Jia, Z.; Ben Amar, M.; Chhor, K.; Brinza, O.; Abderrabba, M.; Vignes, J.-L.; Kanaev, A. Nanoparticulate TiO_2-Al_2O_3 Photocatalytic Media: Effect of Particle Size and Polymorphism on Photocatalytic Activity. *ACS Catal*. 2012, 2, 1884–1892, doi: 10.1021/cs300033y.

[4] Jia, Z.; Vega-Gonzalez, A.; Amar, M. Ben; Hassouni, K.; Tieng, S.; Touchard, S.; Kanaev, A.; Duten, X. Acetaldehyde removal using a diphasic process coupling a silver-based nano-structured catalyst and a plasma at atmospheric pressure. *Catal. Today* 2013, 208, 82–89, doi: http://dx.doi.org/10.1016/j.cattod.2012.10.028.

[5] Sauce, S.; Vega-González, A.; Jia, Z.; Touchard, S.; Hassouni, K.; Kanaev, A.; Duten, X. New insights in understanding plasma-catalysis reaction pathways: study of the catalytic ozonation of an acetaldehyde saturated $Ag/TiO_2/SiO_2$ catalyst. *Eur. Phys. J. Appl. Phys*. 2015, 71, 20805, doi: 10.1051/epjap/2015150020.

[6] Jia, Z.; Barakat, C.; Dong, B.; Rousseau, A. VOCs Destruction by Plasma Catalyst Coupling Using AL-KO PURE Air Purifier on

Industrial Scale. *J. Mater. Sci. Chem. Eng.* 2015，3，19.

[7] Jia，Z.；Rousseau，A. Sorbent track：Quantitative monitoring of adsorbed VOCs under in-situ plasma exposure. *Sci. Rep.* 2016，6，31888，doi：10.1038/srep31888.

[8] Jia，Z.；Wang，X.；Thevenet，F.；Rousseau，A. Dynamic probing of plasma-catalytic surface processes：Oxidation of toluene on CeO_2. *Plasma Process. Polym.* 2017，14，1600114.

[9] Jia，Z.；Wang，X.；Foucher，E.；Thevenet，F.；Rousseau，A. Plasma-catalytic mineralization of toluene adsorbed on ceo<inf>2</inf>. *Catalysts* 2018，8，303，doi：10.3390/catal8080303.

[10] Jia，Z.；Ben Amar，M.；Yang，D.；Brinza，O.；Kanaev，A.；Duten，X.；Vega-González，A. Plasma catalysis application of gold nanoparticles for acetaldehyde decomposition. *Chem. Eng.* J. 2018，347，913–922，doi：10.1016/j.cej.2018.04.106.

[11] Lee，H.；Lee，D.-H.；Song，Y.-H.；Choi，W. C.；Park，Y.-K.；Kim，D. H. Synergistic effect of non-thermal plasma–catalysis hybrid system on methane complete oxidation over Pd-based catalysts. *Chem. Eng.* J. 2015，259，761–770，doi：10.1016/J.CEJ.2014.07.128.

[12] Jia，Z.；Ben Amar，M.；Brinza，O.；Astafiev，A.；Nadtochenko，V.；Evlyukhin，A. B.；Chichkov，B. N.；Duten，X.；Kanaev，A. Growth of Silver Nanoclusters on Monolayer Nanoparticulate Titanium-oxo-alkoxy Coatings. *J. Phys. Chem.* C 2012，116，17239–17247，doi：10.1021/jp303356y.

[13] Labidi，S.；Jia，Z.；Amar，M. Ben；Chhor，K.；Kanaev，A. Nucleation and growth kinetics of zirconium-oxo-alkoxy nanoparticles. *Phys. Chem. Chem. Phys.* 2015，17，2651–2659，doi：10.1039/C4CP05149A.

作者简介

Zixian Jia（贾子先），Assistant Professor，University of Paris 13，France

+33 6.26.23.04.61 zixian.jia@gmail.com

COMPETENCES

—Development of nanomaterials using the sol gel and photo chemistry process

—Study of the performance of catalysts in the plasma-catalytic process for depollution

—Characterization：HPLC，DLS，DRX，AFM，MEB，MET，GC-MS，FT-IR，DRIFT

—Language：Chinese，English，French

—Software：Matlab，Origin，Image，Bureautique（Office，Word，Excel，PPT….）

FORMATION

Pre-doctoral formation

2009—2010 Master 2 *Process Engineering*，Institut Galilée-Université Paris 13，France.

2008—2009 Master 1 *Environmental Engineering*，Université de Haute-Alsace，France.

2007—2008 DUFEE Diplôme d'Université de Français pour Etudiants Etrangers，Université de Haute-Alsace，France.

2003—2007 Licence *Chemistry*，Institut de Chimie，Université de Xiamen，CHINE

PhD Thesis （2010—2013）

Title：Development of Ag，Au / TiO2 nanostructured composite materials for the decontamination of gaseous effluents with plasma activation.

Laboratoire des Sciences des Procédés et des Matériaux （LSPM），University of Paris 13，France

PROFESSIONAL EXPERIENCE

09/2011—09/2014 Moniteur in Process Engineering，158H TP，Galilee Institute-University Paris 13，France.

01/2014—07/2017 Postdoc，Laboratoire de Physique des Plasmas，Ecole polytechnique，France

10/2017—08/2018 Postdoc，Laboratoire de Physique des Lasers，University of Paris 13，France

ADMINISTRATIVE RESPONSIBILITIES

2016— Vice President of the Association of Chinese Scientists and Engineers in France （ASICEF）

2015— President of the Alumni Association of the University of Xiamen in France

2011—2014 Member of the Board of the Chinese Doctoral Union in France-Vice-President （2011—2013）—President （2013—2014）.

HONOR

Chinese Academic Excellence Award （2013）

Bronze Medal：Sino-French Competition of Entrepreneurship and Innovation 2013（Project leader）

Nouvelle Perspective dans le Traitement du Glaucome A Novel Perspective in Glaucoma Therapy

Jian Zhang（张剑）

Ancien service d'ophtalmologie，Hôpital Saint-Antoine，Faculté de médecine Pierre et Marie Curie

Résumé：Actuellement，le traitement d'un glaucome primitif à angle ouvert；avec une pression intraoculaire élevée est la chirurgie par trabéculectomie. Mais il existe toujours un fort taux de complication et il persiste des cas réfractaires.Les traitements chirurgicaux traditionnels sont considérés comme des filtrations，alors qu'il serait plus correct，physiologiquement et anatomiquement，de les comparer à des fistules.

Dans le futur，il sera possible d' utiliser des techniques issues de la nanotechnologie et des technologies de cellules souches，ou de la médecine régénérative pour le traitement des glaucomes.

Mots-clés：Glaucome；Humeur aqueuse；Pression intraoculaire；Trabéculectomie；Nanotechnologie；Cellules souches；Médecine；Régénérative.

Abstract:

The treatment of chronic open-angle glaucoma with high intra-ocular pressure begins habitually with eye drops to reduce intra-ocular pressure. When these medicaments are not effective enough and the field of vision is reduced progressively, the antiglaucoma surgery of trabeculectomy can be proposed. This is called filtering operation. However, on term of anatomy and physiology, that actually build only a fistula. Refractory glaucoma also needs repeated surgeries, because of its various complications.

In the future, it will be possible to apply new techniques issues from recent development in nanotechnology and stem cells and regenerative medicine to repair the blocked pathological trabeculum meshwork with high intra-ocular pressure.

Keywords:

Glaucoma; Aqueous humor; Intra-ocular pressure; Trabeculectomy; Nanotechnology; Stem cells; Regenerative medicin.

Pour former une bonne image, le globe oculaire doit conserver sa forme et donc, maintenir une tension intraoculaire entre 10 à 20 mmHg (moyenne 16 mmHg).

Anatomiquement, le globe oculaire est un organe des plus perfectionnés, semblable à une optique automatique. C'est une sphère à grand axe ovoïde en coupe sagittale, dont l'ensemble a une puissance de +58,64 dioptries pour l'œil emmétrope. Son volume est évalué à 6, 5 cm^3 pour un poids de 7 grammes. Le diamètre du centre de la cornée à la fovéa est, en moyenne, de 24 mm, pour une hauteur de 23 mm et une largeur de 23, 5 mm.

L'œil est constitué de trois tuniques ou enveloppes (Figure 1) :

1. La sclérotique ou sclère est la plus externe des tuniques. Elle entoure les 4/5e postérieurs du globe oculaire. Cette membrane, fibreuse et inextensible, est la plus solide et la plus résistante des structures de l'œil et permet d'en assurer la protection tout en servant d'insertion aux muscles oculomoteurs.

La sclérotique se continue en avant par la cornée. Elle est transparente, avasculaire, très innervée et donc très sensible. Elle joue un rôle majeur en apportant à l'œil une puissance est de +43 dioptries (son rayon de courbure de la face avant est de 7, 8 mm, celui de la face arrière est de 6, 8 mm). La cornée a une épaisseur variable: en son centre, l'épaisseur varie entre 0, 4 et 0, 5mm, en périphérie, elle mesure entre 0, 9 et 1mm. L'indice de réfraction de la cornée est de 1, 377 et son poids est de 1, 2 g en moyenne.

La circonférence équatoriale de la sclérotique est de 77 mm. Son épaisseur est variable selon les régions: de 0, 6 à 0, 8 mm au niveau des limbes, 0, 5 mm à l'équateur, 0, 3 mm en arrière de l'insertion des muscles droits et 1 mm au voisinage du nerf optique.

2. L'uvée (tunique intermédiaire). C'est une couche située entre la sclérotique et la rétine. Ce tissu richement vascularisé nourrit l'iris et la rétine. Elle est formée en arrière par la choroïde qui se prolonge en avant par le corps ciliaire et l'iris.

L'iris donne la couleur à l'œil et contrôle la dilatation de la pupille grâce à ses propriétés contractiles: c'est un diaphragme s'adaptant à la quantité de lumière reçue: quand le diamètre est petit, la profondeur de champ augmente alors que, dans des conditions de faible luminosité (la nuit), la pupille se dilate et l'image qui se forme sur la rétine est moins nette.

La choroïde est formée de nombreux pigments colorés et constituant un écran à l'intérieur de l'œil réalisant comme une chambre noire.

3. La rétine (tunique interne). C'est la tunique sensorielle. Elle est formée d'un ensemble de fibres qui se rassemblent pour former le nerf optique. Le nerf optique rassemble toutes les fibres optiques de l'œil et transmet l'information visuelle de l'œil au cerveau, en particulier à la région postérieure du cerveau où se trouve la région de la reconnaissance visuelle.

Trois milieux transparents remplissent la cavité oculaire:

1. L'humeur aqueuse. Elle contenue dans la chambre antérieure et la chambre postérieure de, respectivement, 0, 25ml et 0, 06ml de volume.

La chambre antérieure est délimitée par la cornée et l'iris antérieur. La jonction de la cornée et de l'iris forme l'angle iridocornéen. L'iris postérieur, le cristallin et le corps ciliaire délimitent la chambre postérieure.

L'humeur aqueuse est sécrétée par le corps ciliaire dans la chambre postérieure, traverse la pupille pour occuper la chambre antérieure. C'est un liquide nourricier limpide comme l'eau, situé dans l'espace entre le cristallin et la cornée (segment antérieur). L'humeur aqueuse a un rôle optique et métabolique nutritionnel vis-à-vis de l'endothélium cornéen (couche unicellulaire formée de 500 000 cellules hexagonales) et de l'épithélium avasculaire cristallinien de la capsule antérieure. Son Indice de réfraction est de 1, 33.

2. Le cristallin et la zonule. Le cristallin est une lentille biconvexe dont les faces antérieure et postérieure se réunissent au niveau de l'équateur. Il est entouré d'une capsule et est relié au corps ciliaire par la zonule de Zinn. Normalement, il est transparent, sans vascularisation ni innervation. Les fibres cristalliniennes sont produites en périphérie par une couche de cellules antérieures qui apportent également les éléments nutritifs au cristallin dans sa portion antérieure.

Le cristallin est composé, à l'âge adulte, de plusieurs couches de fibres cristalliniennes disposées à la manière de pelures d'oignon autour d'un noyau embryonnaire (formation en grain de café) et d'un noyau fœtal. Son poids est d'environ 200mg. L'indice du cristallin est de 1, 42, sa puissance de +16 dioptries sans accommodation.

La capsule du cristallin est une membrane élastique solide, épaisse de 13μm en avant et de 4μm en arrière. Le cristallin, en se bombant, est capable d'augmenter sa puissance: c'est l'accommodation. Avec l'âge, l'accommodation diminue: c'est la presbytie, elle survient autour de 45 ans. A 60-70 ans, l'accommodation est proche de 0, reflet d'une perte presque complète de l'élasticité du cristallin.

3. Le corps vitré. La forme du corps vitré est grossièrement sphérique. Son volume, de 4cl, constitue les 2/3 du volume du globe. Il contient environ 90% d'eau et son poids est de 4 g. Son indice de réfraction est de 1, 334. Le corps vitré est une substance gélatineuse transparente remplissant la cavité vitréenne située en arrière du cristallin.

Le corps vitré présente des adhérences avec les structures oculaires, en avant avec la face postérieure du cristallin par le ligament de Wieger, mais son adhésion la plus solide s'effectue à la base du corps vitré en s'étendant de la moitié de la pars plana à la partie antérieure de la périphérie rétinienne.

En arrière, l'adhérence se fait autour de la papille, autour de la macula et au niveau des vaisseaux rétiniens. En périphérie le corps vitré est condensé et constitue le cortex du corps vitré, appelé membrane hyaloïde antérieure et postérieure.

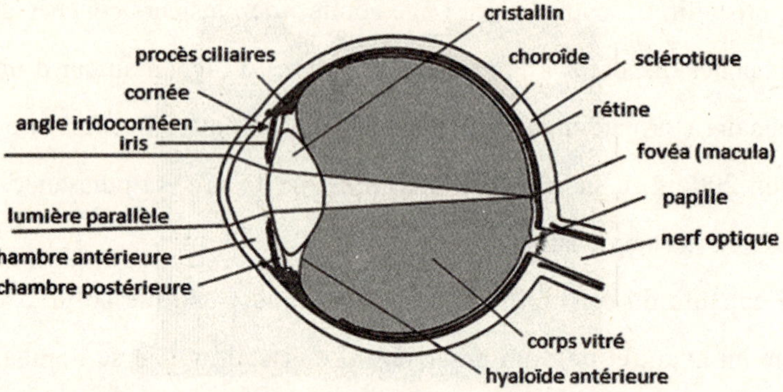

Figure 1 Coupe schématique sagittale d'un globe oculaire，montrant les trois tuniques de l'œil (sclérotique et cornée，uvée，rétine) et trois milieux transparents remplissant la cavité oculaire（humeur aqueuse，cristallin et corps vitré）

A l'âge de 7 ans，le globe oculaire a atteint son poids et son volume adulte，qui ne varieront quasiment plus.

Le maintien de la tension intraoculaire se fait uniquement par l'équilibre entre sécrétion et élimination de l'humeur aqueuse，les deux autres milieux de la cavité oculaire (le cristallin et le corps vitré) étant quasiment fixe.

Le schéma ci-dessous (Figure 2) décrit le trajet de l'humeur acqueuse dans la chambre postérieure et la chambre antérieure. L'humeur aqueuse est produite par les corps ciliaires et est évacuée au travers du réseau trabéculaire，puis du canal de Schlemm et du plexus veineux épiscléral (flèches blanches sur le schéma). Des voies d'excrétion uvéosclérales ont été récemment découvertes (flèche noire sur le schéma)

L'écoulement de l'humeur aqueuse varie en fonction de l'âge et de l'état oculaire.

Le trabéculum est généralement divisé en trois parties. La portion uvéale

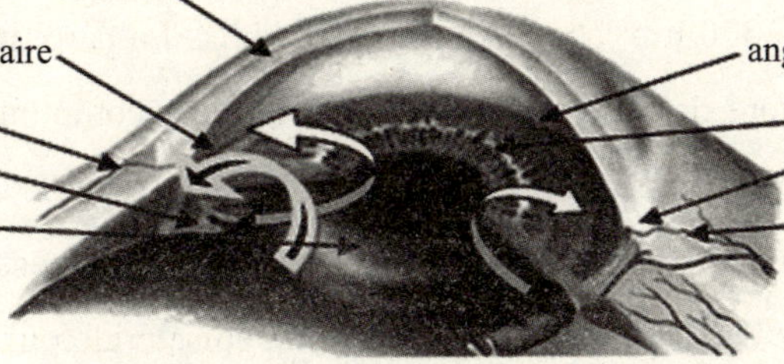

Figure 2 Flux aqueux normal. Coupe shématique du segment antérieur de l' oeil normal.

est adjacente à la chambre antérieure et s'étend, en bande, de la racine de l'iris et des corps ciliaires jusqu'à la cornée périphérique.

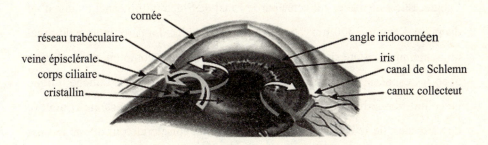

Le trajet de l'humeur aqueuse: l'humeur aqueuse est formée par les procès ciliaires, qui sont composés d'une double couche épithéliale et d'un stroma fortement vascularisé par des capillaires fenêtrés. Chaque procès ciliaires, au nombre de 80 ou plus, est vascularisé par un grand nombre de capillaires, principalement issus des branches du grand cercle artériel de l'iris. Sa formation et sa sécrétion dans la chambre postérieure résultent d'une sécrétion active qui se déroule dans la double couche de l'épithélium ciliaire (la couche cellulaire épithéliale non pigmentée externe et la couche pigmentée interne ainsi que leurs jonctions cellulaires étroites constituent une composante importante de la barrière hémato-aqueuse), de l'ultrafiltration et d'une diffusion simple. Le flux de l'humeur aqueuse est estimé à 2, 0 à 2, 5μl/min, permettant un taux de renouvellement de 1% de l'humeur aqueuse de la chambre antérieure toutes les minutes (le volume de la chambre antérieure étant de 200μl à 250μl). L'humeur aqueuse traverse ensuite les fibres de la zonule, passe en avant du cristallin, s'engage dans la pupille pour arriver dans la chambre antérieure, en avant de l'iris, pour être résorbée dans l'angle irido-cornéen au travers du trabéculum (environ 200 000 à 300 000 cellules trabéculaires par œil) situé au fond de l'angle. Elle est ensuite collectée par le canal de Shlemm (un canal unique d'un diamètre d'environ 370 μm, traversé par des tubules. La paroi interne du canal de Schlemm contient des vacuoles géants qui ont une communication directe avec les espaces intertrabéculaires. La paroi externe est une couche unique de cellules endothéliales qui ne contient pas de pores). Un système complexe de vaisseaux lie le canal de Schlemm à la veine épisclérale qui draine le sang jusqu'a la veine ciliaire antérieure et ophtalmique supérieure, pour se drainer, enfin, dans le sinus caverneux.

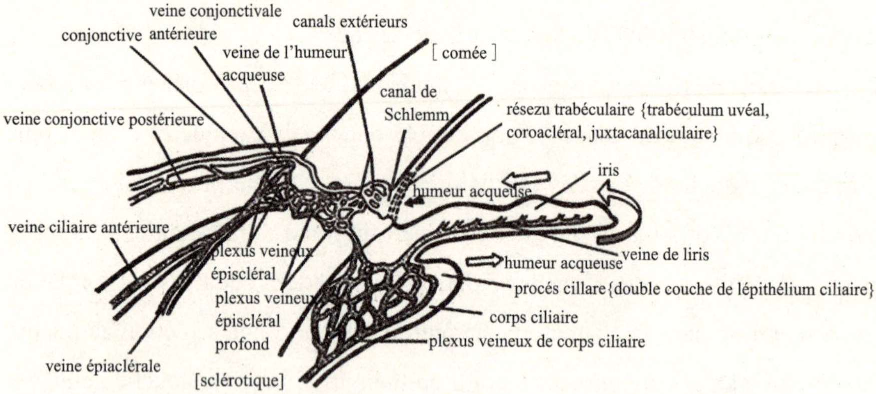

Figure 3 Coupe schématique sagittale du segment antérieur de l'oeil normal, montrant les différentes structures impliquées dans la formation et la résorption de l'humeur aqueuse. L'humeur aqueuse est sécrétée par les procès ciliaires dans la chambre postérieure puis passe dans la chambre antérieure au travers de la pupille. La majeure partie s'écoule dans le système veineux au travers du trabéculum et du canal de Schlemm. Plexus veineux épiscléral→veine épisclérale→veine ciliaire antérieure→veine ophtalmique supérieure→ dans le sinus caverneux.

(d'après ZHANG Jian。Le canal de Schlemm.Chinese Journal of Ophthalmology。1979; 25: 125-128)

Le glaucome: le terme *glaucome* regroupe un ensemble de maladies qui ont en commun une *neuropathie optique* caractéristique associée à une perte de la *fonction visuelle*. L'élévation de la pression *intraoculaire* (PIO) est un des premiers facteurs de risque. Généralement, une PIO élevée est liée à une augmentation de la résistance à l'excrétion de l'humeur aqueuse. La résistance à l'évacuation de l'humeur aqueuse siège au niveau du réseau trabéculaire et du canal de Schlemm; on pense que le site spécifique de la résistance à l'écoulement de l'humeur aqueuse est situé au niveau du trabéculum juxtacanaliculaire. Le trabéculum juxtacanaliculaire est adjacent au canal de Schlemm, dont il forme la paroi interne(Figure 3).

Actuellement, le traitement d'un glaucome primitif à angle ouvert avec une PIO élevée est la chirurgie par trabéculectomie. Mais il existe toujours un fort taux de complications et il persiste des cas réfractaires. Les traitements chirurgicaux traditionnels sont considérés comme des filtrations, alors qu'il serait plus correct physiologiquement et anatomiquement de les comparer à des fistules.

Dans le futur, il sera possible d'utiliser des techniques issues de la nanotechnologie et des technologies de cellules souches ou de la médecine régénérative pour le traitement de glaucomes. En effet, ces techniques ont d'ores et déjà été largement étudiées pour réaliser des tissus fonctionnels tels que les tissus de cardiomyocytes, des cornées, etc., et constitueraient une voie à développer dans le traitement du glaucome avec, pour objectif de reconstruire les voies d'évacuation de l'humeur aqueuse. Ceci étant, on pourrait utiliser les cellules souches pluripotentes induites (CSPi) humaines pour réaliser le canal de Schlemm in vitro puis l'implanter dans l'œil du patient.

Références

[1] Encyclopédie médico-chirurgicale (EMC ophtalmologie) [21–280–B–50].Traitement chirurgical du glaucome à angle ouvert.

[2] Duke-Elder WS，et al：System of Ophthalmology，1961，Vol II，186–215，Kimptom，London.

[3] ZHANG Jian. Le canal de Schlemm. Chinese Journal of Ophthalmology. 1979，25：125–128.

[4] Textbook of Glaucoma.3rd ed. Baltimore：Williams & Wikins；1992.

张剑简介

1946年12月19日出生在浙江省温州市龙湾区（原永强镇永乐、永兴村）。

1964年高中毕业于温州勤俭（原永嘉）中学。同年考入浙江大学医学院，医疗系（原浙江医科大学）

1969年毕业，浙江医科大学医疗系本科文凭。

1970年7月31日，在浙江省军区服兵役至1971年12月28日。

1971年12月28日，由国家分配在温州医学院眼科教研组任教并任附属第一医院眼科、眼外科专科医师。师从中国极负盛名的眼科权威缪天荣教授。

重要论文：1974年，他在温州医学院《温州医药》（眼科专辑）开始发表论文，其中在一九七四年十二月，第二期（总7期）上，《浅谈双眼立体视觉及其测定》一文，日后被广泛引用。如孟祥成、李俊洙编著的《斜视弱视学》1982年，黑龙江人民出版社。徐广第编著的《眼屈光学》1978年，上海科学技术出版社和1995年由军事医学科学出版社再版发行的修订版《眼科屈光学》等。

1979年他发表在“中华眼科杂志”1979年第15卷第2期，125～128页上的《关于“巩膜静脉窦”这一解剖名词的商榷》，全面阐明了“眼球房水排出系统”中Schlemm's canal，这条重要“通道”……得到了我国眼科界和解剖界同仁的认同，并纠正了以往沿用的、不科学的眼球解剖学中“巩膜静脉窦”这个名称 。

1986年5月4—10日，他作为法国眼科协会代表，出席在罗马召开的“第25

届世界眼科大会”，与我中国眼科导师缪天荣在大会宣读《对数视力表和5分记录法》，全文刊登在“大会文献集”『《LOGARITHMIC VISUAL ACUITY CHART AND 5-GRADE NOTION》TIAN-RONG MIAO (1) and JIAN CHARG (2)，P846-854. ACTA XXV CONCILIUM OPHTHALMOLOGIEUM. Proceedings of the XXVth International Congress of Ophthalmology，held in Rome，Italy，May 4-10，1986』(1) Department of Ophthalmology，Wenzhou Medical College，Wenzhiu，Zhejiang，People's Republic of China；(2) Service d'Ophtalmologie de l'Hôpital Saint-Antoine（Professeur H. Saraux），184，rue du Faubourg Sant-Antoine. F-75012 Paris，France

1990年《标准对数视力表》被制定为国家标准（GB11533-89），并在全国实施至今。

1978年他在中国“文化大革命”结束后，恢复研究生制度的第一年，考取了当时中国眼科领域首次设立的“眼科视光学”5位硕士研究生之一。

1983年9月他赴法国巴黎“皮埃尔和玛丽·居里大学医学院（又称巴黎第六大学医学院）”留学。有幸成为世界和法国著名眼科老前辈亨利·萨霍（Henry SARAUX）教授门下深造的第一位中国眼科专业的学生。先后获得“眼科外籍主治医师职称”和医学博士、眼科专家学位。自1988年开始在巴黎第六大学医学院（LA FACULTÉ DE MÉDECINE，UNIVERSITÉ PIERRE ET. MARIE. CURIE Paris VI）从事临床、教学和科研工作。

1986开始，他在“法国国家健康与医学研究院”『（INSERM）Adeministration Déléguée Paris St-Antoine』参加“眼人工角膜”科研项目。随后致力于“开角型青光眼房水排出要道小梁和Schlemm's canal”的研究。研究方向是：人体器官修复、再生技术，微流控技术，研制组成Schlemm氏管内壁、单层内皮细胞的诱导多功能干细胞，甚至人工器官的技术等，推动攻克青光眼治疗、特别是针对随着年龄的增长而发生“小梁和Schlemm's canal组织结构老化所致的房水排出受阻眼内压升高”的难题。

张剑论文：《青光眼治疗的新思路》摘要

慢性开角型青光眼【Chronic open-angle glaucoma（英语）Glaucome primitif à angle ouvert（法语）】作为高眼压致盲中的一种主要眼疾，是目前全球难以治愈的、世界第二大致盲眼病。青光眼是一种多致病因素疾病，它以视网膜神经节细胞受累（丢失）为特点，导致特征性视神经损害及视野缺损。其中高眼（内）压是青光眼发生和发展的主要危险因素，降低眼内压是青光眼的主要治疗方法（从眼药水至各种引流手术）。本文旨在简要介绍人体眼球前部解剖中被称为眼后房和眼前房，这个仅容纳约2.5～0.3毫升眼房水（它的循环半流期约45分钟），发生病理性循环的机制。被称为房水排出系统的前房角，其中网状小梁结构和房水从眼球内排出管道“Schlemm氏管”内壁衔接处，是房水排出障碍引起眼球内压升高重要原因之一。作者认为，二十一世纪迅速发展的“人类的认知极限、宏观—微观—超微观世界的关系等跨学界和前卫的科学”，可以从人体器官修复、再生技术，微流控技术，研制组成Schlemm氏管内壁、单层内皮细胞的干细胞，甚至人工器官的技术等，来推动直至最后攻克青光眼的治愈。【提供论文的时间是：2014年10月30日】

L'image de l'Empereur Kangxi sous la Plume des Jésuites[①]

Rui Sang

School of Foreign Languages, North China Electric Power University

Beijing 102206, China

Résumé: Kangxi est non seulement un souverain remarquable de la dynastie des Qing, mais également l'un des empereurs les plus brillants de l'histoire de la Chine. Ayant un esprit tolérant, il accepta les Jésuites à la Cour. Grâce aux relations des Jésuites, il devient aussi l'empereur chinois qui suscite la plus grande curiosité en Europe. L'intention de cet article est d'analyser l'image de Kangxi dans les œuvres des Jésuites, et d'interpréter leurs descriptions sur cet empereur. Cet article est essentiellement basé sur les relations des Jésuites français. A l'aide des sources de première main, l'image de ce grand empereur nous paraît de plus en plus claire.

Mots-clefs: Kangxi; image; Jésuites; français.

---

① Supported by "the Fundamental Research Funds for the Central Universities (2019MS069)"

Dans l'édition 2005 du Petit Larousse, on note un article court sur Kangxi: 《*Kangxi*: *Pékin* 1654-id. 1722, *empereur de Chine* (1662–1722) *de la* d*ynastie Qing. Homme de lettres tolérant*, *il accepta des jésuites à sa cour.*》 Ce petit article résume parfaitement les descriptions essentielles des Jésuites sur l'empereur Kangxi: homme de lettres, tolérant, bienveillant avec les Jésuites. Dans cet article, nous cherchons à étudier d'une manière plus poussée la vision des Jésuites sur Kangxi dans plusieurs domaines.

1. **Empereur tolérant et modeste**

Comme le souverain supérieur de l'empire, Kangxi fut souvent loué pour sa tolérance et sa bienveillance. Les Jésuites français nous donnèrent beaucoup d'exemples pour montrer la bonté de l'empereur Kangxi pour son peuple. Selon eux, l'Empereur se souciait toujours de la vie de ses sujets. Il n'hésita pas à prendre des mesures pour aider les gens qui étaient dans le besoin à passer les moments difficiles. Les Jésuites mentionnèrent à plusieurs reprises dans leurs relations que l'Empereur paraissait comme 《le père de son peuple》.

Kangxi menait une vie sobre et simple. Il détestait l'extravagance et les dépenses excessives. 《*Quoyque l'Empereur de la Chine soi sans contredit le Prince du monde le plus puissant*, *soi pour les tresors immenses sont il dispose*, *soit pour l'abondance & la vaste étendue de ses Estats*; *il est extrémement éloigné du luxe dans tout ce qui sert precisement à sa personne*》, écrivit le Père Bouvet. [1] Dans cet extrait, la richesse de l'empire contraste la modestie de Kangxi. Dans le journal du Père Gerbillon, on trouve une description de l'ornement des appartements de Kangxi. D'après Gerbillon, dans la salle du premier appartement, le tapis était 《fort commun》, le trône était 《simple》, la chambre où l'Empereur se reposait était 《extrêmement simple，sans peinture，sans dorure，sans tapisserie》. Quant à l'appartement intérieur，le Père Gerbillon le trouva 《sans avoir rien de grand ni de magnifique》 et 《sans aucunes richesses》. Suivant la description，il conclut que l'Empereur était 《extrêmement réservé par rapport à sa dépense particulière》，bien qu'il fût 《sans contredit le plus riche Prince du monde》. De plus，l'Empereur était assez éloigné du luxe pour lui-même. Il s'habillait ordinairement de manière fort simple，à la mandchoue. Les Jésuites s'étonnèrent parfois la modestie ordinaire de son habillement. On voit bien que l'empereur Kangxi possédait un esprit lucide sur la dépense de l'argent.

2. L'attitude de Kangxi vers les sciences occidentales

Kangxi est bien connu par son goût de l'étude，c'est un empereur lettré et savant. Comme tous les empereurs réputés dans l'histoire de la Chine，il attacha de l'importance à l'étude des Classiques de confucianisme. Cependant，ce qui suscite la plus grande curiosité des Européens dès son époque，c'est le grand intérêt qu'il portait aux sciences occidentales.

Ayant une grande curiosité，l'empereur Kangxi s'intéressait à toutes les disciplines. Il se consacra tout d'abord aux mathématiques. Il choisit lui-même l'arithmétique，les éléments d'Euclide，et la géométrie pratique. Les Jésuites à la Cour，comme le Père Antoine Thomas，le Père Gerbillon et le Père Bouvet，lui servaient de professeurs. Il ne cessa point de leur poser des questions，jusqu'au moment où il pût enfin comprendre leurs raisonnements. En même temps，la musique européenne attira son attention. Il invita le père Pereira à composer un manuel，et ordonna aux artisans à la Cour de fabriquer plusieurs instruments musicaux occidentaux. Selon les écrits des Jésuites，peu de temps après，l'Empereur savait déjà exécuter de petits morceaux de musique.

Kangxi attacha également de l'importance à la médecine occidentale. Il établit un laboratoire de pharmacie à la Cour pour les Jésuites. Pour mieux apprendre la médecine, il demanda aux Jésuites de composer des manuels médicaux. S'intéressant particulièrement au corps de l'homme, il apprit sérieusement l'anatomie. Hormis tout cela, l'empereur Kangxi avait une grande passion pour la géographie, l'astronomie, les langues européennes, la philosophie, etc. Il montrait un intérêt débordant.

Dans les récits des Jésuites, on retrouve l'ardeur et l'assiduité de l'empereur Kangxi pour les études. Pour la commodité des Jésuites, il leur donna un de ses propres appartements pour qu'ils pussent donner leurs cours. Admirant de plus en plus la solidité des sciences occidentales, il n'hésita pas à consacrer de longues heures à apprendre les sciences. Il passa la plupart de ses journées avec les Jésuites. Dans la lettre écrit en 1703 au R. P. de la Chaise, le Père Fontenay décrivit: 《*Ils partaient de Pékin dès quatre heures du matin, et ne revenaient qu'au commencement de la nuit. A peine étaient-ils de retour, qu'il fallait se remettre au travail, et passer souvent une partie de la nuit à composer et à préparer les leçons du lendemain... L'Empereur continua cette étude pendant quatre ou cinq ans...*》[2] Selon les Jésuites, l'Empereur détestait toujours 《la vie de l'oisiveté》.

Kangxi possédait un esprit pragmatique. Quand il apprenait de nouvelles choses, il ne se contenta pas de la seule réflexion. Au contraire, il y joignit la pratique. Selon les Jésuites, l'Empereur s'intéressait à tous les sujets scientifiques sans exception, et plus particulièrement à ceux qui pouvaient être immédiatement utilisés, comme Bouvet le dit: 《*Au reste ce sage Prince sçait profier admirablement de tout ce qui peut servir à la conservation de son Etat.*》[3]

3. **L'attitude de Kangxi vers la religion chrétienne**

Bien que les Jésuites aient éprouvé toutes sortes de difficultés et de souffrances en Chine, ils faisaient toujours preuve de zèle pour l'établissement de la religion chrétienne dans le grand empire. Ils rendirent service à l'Empereur afin d'obtenir sa protection et de l'amener à leur religion. En adoptant cette idée, les Jésuites travaillèrent sérieusement à la Cour, et se dépensèrent sans compter pour aider l'Empereur à apprendre les sciences occidentales. Heureusement, leurs efforts furent récompensés. Dans leurs récits, ils témoignèrent des grandes faveurs dont ils bénéficiaient auprès de l'empereur Kangxi.

A la Cour de Pékin, les Jésuites reçurent de temps en temps des dons de l'Empereur, parmi lesquels la nourriture était la plus fréquente. La première fois quand l'Empereur reçut en audience le groupe des 《mathématiciens du roi》, il les honora du thé et de l'argent qui semblait 《une gratification extraordinaire》 même pour les Chinois. L'Empereur réserva souvent des mets impériaux ou toutes sortes de choses rares de la Cour pour les faire goûter aux Jésuites. Le Père Gerbillon enregistra minutieusement ces évènements dans son journal: 《*L'Empereur ayant envoyé des fruits & des confitures de sa table aux Peres Pereira & Thomas*》《*L'Empereur nous envoya encore des confitures séches de la table, que nous distribuâmes chefs du tribunal.*》《*Sa Majesté nous fit dire après le dîner, qu'il vouloit nous faire goûter du vin qu'on lui avoit envoyé des provinces méridionales*》[4]

De la nourriture aux vêtements, du vin au logement, l'Empereur fit de son mieux pour rendre la vie des Jésuites à la Cour plus confortable. En même temps, il essaya toujours de trancher les difficultés de l'évangélisation des Jésuites. Le Père Gerbillon relata un événement dans son journal: il y avait une persécution contre les Chrétiens dans la Province de Shandong. Pour sauver leurs compagnons, les Jésuites qui restaient à la Cour décidèrent d'aller voir l'Empereur et de lui demander la protection. Ayant entendu la plainte des Jésuites, l'Empereur les aida enfin à calmer cette affaire. La persécution à la Province de Shandong s'arrêta.

Les Jésuites furent émus de la tolérance de Kangxi. Ils ne cachèrent jamais leur joie dans leurs récits. De plus, la faveur de Kangxi fit entrevoir aux Jésuites de grandes espérances. Selon eux, Kangxi fut perméable à l'influence européenne. Ils pensaient même qu'un jour, Kangxi pourrait se faire chrétien, comme le Père Bouvet le dit: 《*les heureuses dispostions...que ce Prince fait paroître à l'égard de la Religion, nous donnent lieu d'augurer, qu'il sera peut-être un jour le destructeur de l'Idolâtrie dans la Chine, pour ressembler de plus prés à Vôtre Majesté, qui a mis sa principale gloire à abattre l'heresie dans ses Etats, et à étendre la Religion par tout le monde.*》 [5] Selon Bouvet, il christianisme deviendrait un jour la 《religion dominante》 de la Chine.

En réalité, cette pensée était trop optimiste. L'empereur Kangxi semblait parfois intéressé à la religion chrétienne, mais c'était uniquement par curiosité. Ayant une passion pour toutes les choses inconnues, il considéra le christianisme comme une religion exotique, même comme une science. Par esprit d'ouverture, il accepta volontiers toutes les religions, pourvu qu'elles ne constituassent une menace pour son autorité absolue. En tant que souverain, il considéra les missionnaires comme ses sujets chinois, et ne voulait en aucun cas les voir se soumettre aux ordres d'un autre. De plus, sa faveur n'était que pour les Jésuites. Quant aux convertis chinois, il n'avait aucun d'affection. Il ne pensa en aucun cas à se convertir. Son attachement aux Jésuites fut sincère mais lucide. En effet, il existait toujours une contradiction inconciliable entre le but de Kangxi et celui des Jésuites.

Les relations des Jésuites sont d'éloquents documents sur Kangxi et sur la Cour mandchoue. Elles firent également écho aux philosophes pendant la période des Lumières. Un empereur tolérant, modeste, assidu, intéressé aux sciences occidentales et même à la religion chrétienne, c'est juste l'image de Kangxi décrite par les Jésuites. Bien qu'ils cherchassent toujours à décrire l'Empereur objectivement, on trouve plusieurs exagérations dans leurs relations où ils parlèrent de Kangxi d'un ton admiratif. D'après eux, Kangxi fut un empereur idéal, voire parfait. En tenant compte du contexte historique, il nous faut analyser les descriptions des Jésuites sur Kangxi d'un point de vue objectif, afin de mieux connaître l'image de cet empereur remarquable.

Références

[1] Joachim Bouvet, *portrait historique de l'empereur de Chine*, Paris, E. Michallet, 1697, P.98.

[2] Jean de Fontenay, *lettre au R. P. de la Chaise dans Lettres édifiantes et curieuses de Chine*, notées par Isabelle et Jean-Louis Vissière, Paris, Garnier, Flammarion, 1979, 126-127.

[3] Joachim Bouvet, *portrait historique de l'empereur de Chine*, Paris, E. Michallet, 1697, P. 190.

[4] Jean-François Gerbillon, *journal compris dans descriptions géographique, historique, chronologique, politique et physique de l'Empire de la Chine et de la Tartarie Chinoise*, la Hare, 1736, 267-277.

[5] Joachim Bouvet, *portrait historique de l'empereur de Chine*, *Paris*, *E. Michallet*, 1697, 6-7.

作者简介

Rui Sang（桑瑞）

Enseignante de français à l'Université d'Énergie Électrique de Chine du Nord, Beijing, Chine

Domaines de recherche

littérature comparée, interactions culturelles et littéraires entre la Chine et l'Europe

Éducation

2012—2017 Doctorat en Lettres Modernes, Université Toulouse-Jean Jaurès, Toulouse, France

2010—2012 Master en Littérature française, Université Paris-Sorbonne, Paris, France

2006—2010 Licence en Langue française, Université Sun Yat-sen, Guangzhou, Chine

Expériences Professionnelles

2018—présent Enseignante de français, Université d'Énergie Électrique de Chine du Nord, Beijing, Chine

2013—2016 Chargée de cours de chinois, Université Toulouse-Jean Jaurès, Toulouse, France

致　谢

本论文集得到10个旅法侨团的友情赞助，名单如下：

法国华侨华人会	主席任俐敏暨全体会员
法国青田同乡会	会长孙少荣暨全体会员
法国法华工商联合会	会长戴安友暨全体会员
法国文成联谊会	会长朱少云暨全体会员
法国温州商会	会长胡镜平暨全体会员
中法服装实业商会	会长胡仁爱暨全体会员
法国浙江商会	会长余例权暨全体会员
法国浙江同乡会	会长高敏鏗暨全体会员
法国华人鞋业协会	会长郑建华暨全体会员
法国华侨华人妇女联合会	会长潘笑黎暨全体会员